Matter at
Low Temperatures

P.V.E. McClintock, BSc, DPhil, DSc
D.J. Meredith, BSc, PhD
and
J.K. Wigmore, BA, MA, DPhil, FInstP

Department of Physics,
University of Lancaster

Blackie

Glasgow and London

Blackie & Son Limited,
Bishopbriggs, Glasgow G64 2NZ

Furnival House, 14–18 High Holborn, London WC1V 6BX

© 1984 Blackie & Son Ltd
First published 1984

British Library Cataloguing in Publication Data

McClintock, P.V.E.
   Matter at low temperatures.
   1. Low temperatures   2. Matter—Thermal
properties
   I. Title   II. Meredith, D.J.
   III. Wigmore, J.K.
   530.4   QC278

   ISBN 0-216-91594-5
   ISBN 0-216-91593-7 Pbk

Photosetting by Thomson Press India Ltd.
Printed in Great Britain by Bell and Bain Ltd., Glasgow

Preface

Our aim in writing this book has been to meet the need for a clear and unified introduction to the physics of matter at low temperatures.

We discuss the diverse and fascinating phenomena that occur under these conditions, a number of which have no analogue at all in the everyday world at room temperature, and we indicate the fundamental significance that many of them carry for basic physics. The Third Law of Thermodynamics is treated explicitly only in the Introduction, but it represents an underlying theme implicit in most of what follows. We explain a number of the techniques that are used for reaching low temperatures and we discuss how experimental measurements can be made there in practice. Finally, we present a brief account of the more important of the many applications of low temperatures that are now being realized.

The origins of the book lie in lecture courses offered at the University of Lancaster. Material has been drawn, in particular, from a final year undergraduate (optional) course of the same title and from lectures given to students enrolled for the MSc in Applied Cryophysics. We hope that the resultant synthesis will prove to be of value, not only to students on both sides of the undergraduate/graduate interface, but also to qualified scientists whose interests have previously been concentrated in other areas but who, for personal or professional reasons, find that they now wish to know more about the physics of low temperatures.

We have assumed that the reader has had some prior exposure to undergraduate thermodynamics, quantum mechanics, solid state physics and statistical mechanics and, in particular, that he or she has already encountered the Bose–Einstein and Fermi–Dirac distribution functions. We do not expect, however, that all of this material will necessarily have been fully retained and numerous reminders and signposts are therefore provided at appropriate points in the text. The emphasis throughout is placed on physical concepts rather than on mathematical detail.

The Bibliography at the end of each chapter is divided into two sections: References, and Further Reading. This division is perhaps somewhat arbitrary, and at times inevitably becomes blurred. The intention, however, is that the References should provide an opportunity for the student to look up further details of particular measurements or calculations and to pick up some of the flavour and emphasis of the original research: they are not intended to provide a justification for every statement made in the text and neither should they be construed as an attempt to assess priority of discovery. The Further Reading mostly consists of books and review articles to which the student can turn for general background, a fuller discussion and a more inclusive bibliography of the topic in question. Citations given in the text of a particular chapter may refer to either section of its Bibliography.

This book would probably not have been written were it not for the stimulating environment provided by a lively low temperature research laboratory, and we freely acknowledge our very substantial indebtedness to our colleagues. Our grateful thanks are due to Katherine Bevington, Valerie Wigmore and Pat Fawley who typed our rough manuscript, to Isobel Matthews who prepared many of the figures, and to Tony Guénault and Marion McClintock who were kind enough to read and comment in detail on the initial version of the typescript, and to Marion McClintock again for preparing the index. We are, of course, entirely responsible for any errors of fact or emphasis that remain.

University of Lancaster   P.V.E. McClintock
                                                          D.J. Meredith
                                                          J.K. Wigmore

Contents

1 Introduction

1.1 The significance of low temperatures

Currently-available technology enables man to cool matter (copper nuclei) to 5×10^{-8} K or to heat it (plasma in a tokamak) to 10^8 K. Compared to these extremes of temperature, separated by a factor of more than 10^{15}—which can only increase with the passage of time, as further progress is made—everyday life and work are restricted to a narrow range indeed, as indicated in Fig. 1.1.

It is scarcely surprising that, as the temperature is varied within the huge range available, the properties of all materials undergo very considerable modifications. Many of these changes are of profound significance not only in terms of basic physics, but also in relation to practical applications; and this is as true of temperatures below ambient as it is of those above. The pages which follow are intended to provide an introduction to the physics and technology of the *lower* temperatures found on moving downwards from our starting position just above the middle of Fig. 1.1.

The use of a logarithmic temperature scale for Fig. 1.1 may seem a little arbitrary; but there are a number of excellent reasons for this particular choice. First, it will be noted that the definition of the absolute (Kelvin) scale of temperature is itself made in terms of ratios (the ratio of two temperatures formally being defined as the ratio of the heats accepted and rejected by an ideal engine operating between thermal reservoirs at those temperatures). Secondly, it is a matter of experimental observation and experience that a given change in properties often seems to be related more to the *factor* by which the absolute temperature is varied than to the actual magnitude of the variation. Thus a change of 2.7 K in going from 0.3 K to 3.0 K would be expected, on average, to bring about as large a change in properties as would result from a temperature change of 2700 K between 300 K and 3000 K. Thirdly, a logarithmic plot serves to emphasize the unattainability of the absolute zero of temperature, as is required by the Third Law of Thermodynamics (§1.2), since $T = 0$ would be situated an infinite distance downwards in a plot such as Fig. 1.1.

The technology of reaching and maintaining very low temperatures is

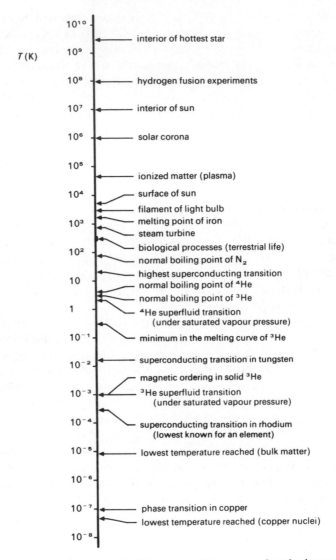

Figure 1.1 The temperatures T (in Kelvin) at which some selected phenomena occur, illustrating the huge range which is accessible. Note that the scale is logarithmic.

referred to as *cryogenics*. It is usually dependent on the use of liquefied gases, known as *cryogens*. This word, formed from the Greek κρύος (cold) and — γενες (generated from) was introduced by Kamerlingh Onnes, who liquefied helium for the first time (at Leiden in 1908) and who can thus be regarded as the progenitor of cryogenics as we know it today. The study of matter at low temperatures is usually called *cryophysics*. Similarly, the prefix 'cryo-' is often coupled to other words to indicate low temperature aspects of the discipline in question (thus, *cryobiology, cryosurgery* and so on). No hard and fast dividing line exists at which ordinary refrigeration becomes cryogenics, and any chosen criterion would clearly be somewhat arbitrary. In the present case, we will not restrict our discussion rigidly to any particular temperature range, although our emphasis will be directed towards the region below, say, 20 K (the boiling point of liquid hydrogen and the approximate temperature below which superconductivity is found) because of its particular interest and importance. Even if we wished to restrict our studies exclusively to this region, however, the properties of materials at higher temperatures would still need to be discussed because they are of crucial importance for the design of *cryostats*, the apparatus used for providing and maintaining the required low temperatures.

An alternative viewpoint should also be mentioned. It is not at all unreasonable to regard the temperature as being 'low' whenever quantum effects in aggregate matter are important. This will be the case in a gas, for example, if the density and temperature are such that the de Broglie wavelength of the constituent particles is greater than their average separation. From this point of view, the electron gas in a metal at room temperature, and the neutrons in a neutron star at 10^8 K, equally constitute low temperature systems and must indeed be treated as such, although gaseous helium under atmospheric pressure at 20 K would not be considered to be at a low temperature. Accordingly, we also include in §6.6 a very brief mention of some super-room-temperature systems which are nonetheless relevant.

It must be emphasized that *temperature* can properly be ascribed only to systems which are in internal equilibrium and that the concept is essentially statistical in nature. Thus, given a large number of atoms in thermal equilibrium, one can in principle determine the temperature T of the whole assembly by measuring the relative number of atoms in a particular excited state of energy E. If the density is not too high or the temperature too low (or if the atoms are localized), so that Boltzmann statistics are applicable, the occupancy of this state relative to all others will on average be proportional to $\exp(-E/k_B T)$, where k_B is Boltzmann's constant. To speak of the temperature of a single isolated atom would, however, be quite meaningless. In what follows, therefore, we will tacitly assume first, that we are dealing with macroscopic quantities of material, for which the statistics will be reliable; and secondly, either that steady state conditions prevail, or that conditions are changing sufficiently slowly (quasistatically) that departures from equilibrium are negligible. It should also be borne in mind that the Kelvin temperature scale is absolute in the sense that it is quite independent of the chosen thermometric property or material, but that the size of the Kelvin degree is arbitrary to the extent that it has been selected so as to give consistency with that of the Celsius scale. Further (detailed and thought-provoking) discussions of temperature, temperature scales and related topics are to be found in, for example, Zemansky and Dittman (1981).

1.2 The Third Law of Thermodynamics

It is a matter of common experience that the addition of heat to a body causes its temperature to increase (provided, of course, that no first-order phase transformations or chemical reactions are involved) and, correspondingly, that the removal of heat causes its temperature to fall. Given the equivalence of heat, work and energy as expressed through the First Law it may, at first sight, seem only natural to conclude that the absolute zero of temperature will be reached when the energy of the body has itself become zero. Further consideration in the light of the predictions of quantum mechanics demonstrates, however, that this cannot be so. In quantum mechanics, the ground-stage energy of a system is seldom zero. For example, a quantum harmonic oscillator (used in §2.2 for modelling lattice vibrations) has energy levels quantized as $(n + \frac{1}{2})h\nu$, where n is an integer, ν is the characteristic frequency and h is Planck's constant. The ground state therefore has the finite energy of $\frac{1}{2}h\nu$. Similarly, in the case of an ideal Fermi–Dirac gas, the ground state energy is not zero but an average of $\frac{3}{5}k_B T_F$ per particle, where T_F is the Fermi characteristic temperature. It is abundantly clear, therefore, that the energy of any given system cannot be assumed to fall to zero at $T = 0$.

Innumerable experimental studies have shown that the quantity which does fall to zero as the absolute zero is approached is in fact the *entropy*, corresponding to the degree of disorder of the system (see below). This experience is embodied in the Third Law of Thermodynamics:

> *The entropy of all systems and of all states of a*
> *system is zero at absolute zero.*

Some qualifications and comments are immediately required. The thermodynamic definition of entropy refers to changes in entropy rather than to its absolute magnitude, the change ΔS in the entropy of a given system when a quantity of heat Q enters it reversibly at temperature T being defined as

$$\Delta S = \frac{Q}{T}. \tag{1.1}$$

Thus, the entropy of a system at absolute zero can in principle be assigned any chosen value, and need not necessarily be zero (although, for convenience, it is normally defined to be zero). The essential point is that all states of the system at absolute zero must have the *same* entropy. That is, the entropy differences between all states of the system which are connected by reversible paths become zero at absolute zero. It must be emphasized, though, that unless such reversible paths do exist, entropy differences between the states in question cannot be defined and the Third Law cannot meaningfully be applied.

It is important to note that the Third Law may be applied separately to quasi-separate subsystems within a system. In discussing a metal, for example, one can treat the phonons and electrons separately insofar as their contributions to the total internal energy and specific heat are concerned. Likewise, one can discuss their separate contributions to the total entropy. Thus, the Third Law can be applied separately to the electron and phonon sub-systems even in those (real) cases where structural defects become 'frozen' into the crystal and consequently persist down to the lowest attainable temperatures. Such defects represent non-equilibrium thermodynamic states, from which reversible paths do not exist. The configurational entropy of such a crystal at absolute zero must therefore be ill-defined; and so the Third Law cannot meaningfully be applied to that particular component of the total entropy.

As mentioned above, there is a close correspondence between entropy and the degree of disorder of a system. The statistical interpretation of entropy is given by the Boltzmann–Planck equation

$$S = k_B \ln \Omega \qquad (1.2)$$

where Ω is the number of accessible microstates (distinguishable microscopic configurations) of the system. Taking, as a first example, a perfect crystal at absolute zero, it is evident that only one microstate is possible: any deviation from the perfect order of the crystal would require additional energy, and this is not available. Thus, $\Omega = 1$, and $S = 0$ in conformity with the Third Law. At finite temperatures, however, the situation is radically different. There may then be vacancies in the lattice and perhaps some interstitial atoms, and there will certainly be phonons (quantized vibrational modes of the lattice: see §2.2). All of these correspond to an increase in the disorder of the crystal. Their presence also implies an enormous increase in Ω, and hence also in S, because energy can then be conserved while being exchanged between the numerous possible excited states, corresponding to the different microstates of the system as a whole.

Another, perhaps even more striking, example of the correspondence of entropy and disorder occurs during the isothermal magnetization of a paramagnetic crystal (§7.4). With zero (or negligible) applied magnetic field, the orientations of the atomic magnets are totally disordered and the number of microstates Ω_{mag} and corresponding entropy S_{mag} of the magnetic subsystem are large. When the magnetic field is increased, however, the atomic magnets start to align along the field direction and there is a corresponding decrease in magnetic entropy. Experimentally, one finds that a considerable amount of heat must be removed during magnetization in order to prevent an increase in temperature. In a sufficiently strong magnetic field, all the atomic magnets will be aligned in the same direction, giving perfect magnetic order, with $\Omega_{mag} = 1$ and $S_{mag} = 0$.

The implications of the Third Law for low-temperature measurements are of particular importance. Here, we will consider as examples the cases of the low-temperature specific heats and coefficient of thermal expansivity. A number of other instances will also appear in the chapters which follow. The entropy $S(T)$ of a system of fixed volume at temperature T can be found by integrating $(dQ/T)_{rev} = C_V dT/T$ between absolute zero and T, so that

$$S(T) = \int_0^T C_V \, dT/T \qquad (1.3)$$

where C_V is the specific heat at constant volume. If C_V remained finite down to $T = 0$, the integral would diverge as $T \to 0$ in direct contradiction to the Third Law and we are forced to conclude, therefore, that

$$C_V \to 0 \quad \text{as} \quad T \to 0. \qquad (1.4)$$

A similar argument may, of course, be applied to the specific heat at constant pressure, C_P. Turning to the thermal expansivity, defined by

$$\beta = \frac{1}{V}\left(\frac{\partial V}{\partial T}\right)_P, \qquad (1.5)$$

we note that, through use of a Maxwell relation, we may write

$$\beta = -\frac{1}{V}\left(\frac{\partial S}{\partial P}\right)_T. \qquad (1.6)$$

The Third Law implies that $(\partial S/\partial P)_T \to 0$ as $T \to 0$, so we conclude that

$$\beta \to 0 \quad \text{as} \quad T \to 0. \qquad (1.7)$$

We shall see that this vanishing of C_V, C_P and β at $T = 0$ is extremely well supported by experimental evidence.

The Third Law can also be stated in a negative form which is of special relevance to the topics which we will be considering:

> *It is impossible to reach the absolute zero of temperature by any finite number of processes.*

The two forms can be shown to be equivalent. Their close relationship can readily be appreciated by considering a set of processes by which one might try to reach the absolute zero. Each of the steps will be assumed either adiabatic or isothermal and, in the interests of maximizing efficiency, we will suppose all of them to be reversible. Figure 1.2 shows entropy S plotted as function of temperature T for two different possible situations described by the parameter x (which would, for example, be the applied magnetic field in the particular case of a magnetic solid, as discussed in §7.4). In (a), the graphs of $S(T)$ come together at $T = 0$ in obedience to the requirements of the Third Law that $S = 0$ at $T = 0$, whereas, in (b), the Third Law is clearly violated.

The putative cooling procedure, considered first in relation to (a), might be as follows: (i) the controlling parameter is changed reversibly and isothermally from x_1 to x_2, thus taking the system from point A to point B at the constant temperature T_1; and (ii) the controlling parameter is changed reversibly and adiabatically from x_2 back to x_1 again, thus taking the system from B to C and

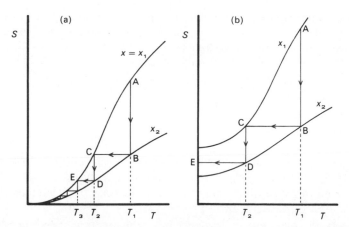

Figure 1.2 Two ways in which the entropy S might vary with temperature T at very low temperatures, for different values of a controlling parameter x. In (a) the entropy, and entropy differences, tend to zero at the absolute zero of temperature, which therefore cannot be reached in any finite number of steps. In (b), where the Third Law is clearly violated, it would in principle be possible to cool the system to absolute zero. In reality (a) always seems to occur rather than (b).

resulting in a final temperature T_2, smaller than the starting temperature. Having established that we can cool material to temperature T_2 we could, in principle, use T_2 as our new starting temperature and repeat the operation, arriving at E with an even lower final temperature, T_3. The procedure could be repeated many times but, because of the convergence of the $S(T)$ curves for x_1 and x_2 and, indeed, those for any other values of x we like to choose, the extent of the cooling will get less each time. On the basis of (a), therefore, we would never quite reach absolute zero in any finite number of operations.

Turning now to (b) of Fig. 1.2, and performing the same sequence of operations, the result is clearly very different. This time, the adiabatic change from D to E will take the system directly to absolute zero. Such a result would be highly gratifying to low-temperature physicists, but it has never occurred in any experiment. In practice, the $S(T)$ curves are invariably found to be like those of (a), in conformity with the Third Law, and absolute zero remains an unattainable goal.

It should be noted that, although the goal cannot be attained, it may nonetheless be approached arbitrarily closely. An interesting comment made by F.E. Simon (1952) is that it must be possible, at least in principle, to reach whatever low temperature may be necessary for the study of any given physical phenomenon. That is, it should be possible to design the experiment in such a way that the entropy changes involved in the phenomenon under study are themselves used to provide the required degree of cooling:

It will only become progressively more difficult to reduce the temperature when the system has lost practically all its entropy. In this case, only a desert lies before us and so we are not interested in going further: we have already to all extents and purposes a system which is hardly distinguishable from one at absolute zero and its properties can be safely extrapolated. If, on the other hand, some new phenomenon is going to occur at lower temperatures, then we will able to make use of it to reach this oasis of interest. Thus the law of unattainability of absolute zero is no barrier to our knowledge.

1.3 Liquefaction of gases

Almost all cryogenic work depends on the provision of liquefied gases. These cryogens have relatively large latent heats of vaporization and they therefore provide convenient constant-temperature baths when boiling under atmospheric pressure. Lower temperatures can, of course, be attained by pumping away the vapour so that the cryogen then boils under reduced pressure. The maximum range of temperature over which any particular cryogen will be directly useful is limited at the top end by the critical temperature T_{cr} (above which the material is no longer a liquid) and, except in the case of helium, at the bottom end by the triple point temperature T_{tp} (below which the vapour pressure of the solid falls rapidly to zero). Relevant physical data for a few selected cryogens are given in Table 1.1. To achieve temperatures below 0.3 K, cryostats incorporating other cooling methods must be used, but even these will normally incorporate baths of, for example: liquid nitrogen (77 K); liquid ^{4}He (4.2 K, or 1.1 K under reduced pressure); and perhaps also liquid ^{3}He (0.3 K under reduced pressure). We consider in this section the physical principles involved in the liquefaction of gases. The ways in which the cryogenic liquids are used in cryostats are discussed in chapter 7, and some of their other practical applications are described briefly in chapter 8.

There are three methods which may, in principle, be employed for the liquefaction of gases: (i) isothermal compression; (ii) the performance of external work by the gas itself; (iii) use of the Joule–Kelvin effect. We will give a brief description of each and will then discuss how combinations of these techniques may be used as the basis of operation of practical liquefiers.

Table 1.1 Physical data for some selected cryogens

Substance	Normal boiling temperature (K)	Critical point		Triple point	
		T_{cr} (K)	P_{cr} (atms)	T_{tp} (K)	P_{tp} (atms)
Ammonia	240	406	112	195	0.060
Oxygen	90.2	155	50.1	54.4	0.0015
Nitrogen	77.3	126	33.5	63.2	0.12
Hydrogen(n)	20.4	33.3	12.8	13.8	0.070
Helium-4	4.21	5.20	2.3	—	—
Helium-3	3.2	3.32	1.2	—	—

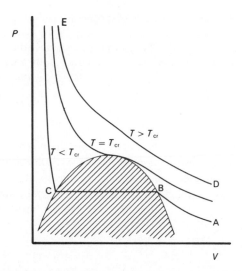

Figure 1.3 Pressure (P)–volume (V) isotherms (schematic) for a hypothetical fluid to illustrate why the vapour cannot be liquefied by isothermal compression unless its temperature T is less than the critical temperature T_{cr}. The shaded area represents the liquid/vapour mixed phase region.

Liquefaction by straightforward compression will only work if the gas is initially below its critical temperature T_{cr}. This may readily be appreciated by inspection of the PV isotherms (Andrews' curves) plotted in Fig. 1.3. For $T < T_{cr}$, an isothermal compression starting from point A meets the gas/vapour coexistence region at B. A further isothermal reduction in volume then takes place at constant pressure (the saturated vapour pressure for the temperature in question) until, at C, the material consists entirely of the liquid phase. Starting from a temperature above T_{cr}, on the other hand, an isotherm such as DE will be followed. In this case there is no phase change and at no stage does the fluid become a liquid, even though its density may increase to a liquid-like value. It should be noted that, of the examples quoted in Table 1.1, only ammonia can in fact be liquefied by the simple application of pressure at room temperature.

If a gas is caused to perform adiabatic external work it must cool, in accordance with the requirements of the First Law of Thermodynamics. What happens is that some of the internal energy of gas, which is largely composed of the kinetic energies of the atoms or molecules, is converted in performing the work. In practice, the gas must first be made to undergo an isothermal compression, during which procedure an amount of heat corresponding to the work being done on the gas will have to be removed; and then it is allowed to expand against a piston or turbine, causing it to cool. The great advantage of this technique is that it invariably results in cooling, whatever the starting conditions. The main disadvantages are, first, that the technique becomes less effective as the temperature falls and the liquefaction point is approached and, secondly, that it is necessary to incorporate moving mechanical components at very low temperatures, which pose obvious problems in terms of lubrication and reliability. The latter difficulties can, however, be overcome by careful design. Because of the decrease in cooling effect with falling temperature it is usual to combine the external work method with another technique, such as Joule–Kelvin expansion (see below).

A variety of different arrangements have been used for expanding the gas and extracting the work. The gas can, for example, be expanded via a turbine, or it can be made to operate a valved piston-in-cylinder arrangement. In either case, it emerges at reduced pressure and temperature, with the work being passed via a shaft or connecting rod to a damping system, usually at room

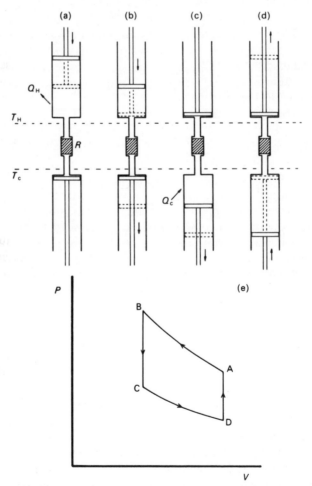

Figure 1.4   The Stirling refrigeration cycle: (a)–(d) represent different positions of the pistons and (e) is the corresponding pressure-volume indicator diagram for the cycle. For description of operation, see text.

temperature. A variant is a Stirling engine operated in reverse (sometimes known as the Kirk cycle), and we will take this as an example for consideration in some detail because of its interest and its importance in practical applications.

The essential features of the Stirling refrigeration cycle are illustrated in Fig. 1.4. Two piston-and-cylinders are used, one situated at the higher temperature end of the system and the other at the colder end. They are connected together via a *regenerator*, a device which has a low thermal conductivity, a small internal volume and a low flow impedance for gas, but which possesses a large heat capacity and a large internal surface area. In describing the cycle, we assume that steady operating conditions have been reached, so that the upper cylinder and the top of the regenerator are both at temperature T_H, whereas the bottom of the regenerator and the lower cylinder are at T_C. The starting position, shown by the full lines of Fig. 1.4(a) and corresponding to point A on the P–V indicator diagram in Fig. 1.4(e), is such that the upper piston is fully opened and the lower one fully closed. If we neglect the relatively small free volume inside the regenerator, all of the gas starts at the hotter end of the system. The cycle then proceeds as follows, where (a)–(d) refer also to Fig. 1.4:

(a) The upper piston moves so as to compress the gas at the hotter end of the system, while the lower piston stays still. Work is being done on the gas and so a quantity of heat Q_H must be removed from it in order to prevent its temperature from rising above T_H. The representative point moves from A to B in Fig. 1.4(e).

(b) Both pistons move together so as to transfer all the gas at constant volume to the colder end of the system, and the representative point moves from B to C. Because of the large heat capacity of the regenerator, the gas is already at T_C and a correspondingly lower pressure when it enters the lower cylinder.

(c) The lower piston moves so as to expand the gas at the colder end of the system, while the upper piston stays still, and the representative point moves from C to D. Work is being done on the piston by the gas and so a quantity of heat Q_C must be absorbed from the surroundings in order to prevent its temperature falling below T_C: this, therefore, represents the refrigeration stroke.

(d) Finally, the gas is returned via the regenerator at constant volume to the hotter end of the system. In doing so, it picks up heat from the regenerator (an amount equal to that deposited there during phase (b)) and enters the upper cylinder at temperature T_H and a pressure equal to its original starting value. Thus, the representative point moves from D back to A again.

The cycle then repeats. The practical realization of the Stirling refrigerator is, of course, a little different from this idealized scheme. It makes for more straightforward engineering if the pistons execute harmonic motions, rather than the discontinuous movements described above; the internal volumes of real regenerators are not, in fact, negligible; and the machines are usually operated at speeds such that phases (a) and (c) are very far from being isothermal. Nonetheless, the underlying principle of operation is as described above, with the net result that heat is absorbed at the colder end of the system.

The third basic technique used in the liquefaction of gases, Joule–Kelvin expansion, depends on exploiting the existence of attractive interatomic forces. It may, in a sense, be regarded as a method of making the gas perform 'internal work' against these forces. The technique would not, therefore, work for an ideal gas where, by definition, the interatomic forces are zero; but, for real gases, cooling can be obtained provided that the starting temperature is less than a certain characteristic value which is different for each gas. The gas is allowed to expand through a porous plug, or an orifice, as sketched in Fig. 1.5(a). Here it is assumed that the gas above the plug is at pressure P_i, temperature T_i; and that, below the plug, the gas is at a lower pressure P_f (and flows at a correspondingly higher velocity). The final temperature T_f is in general different from T_i, but may be either higher or lower. For steady-state conditions, it can be shown (Zemansky and Dittman, 1981) that the molar enthalpy h of the gas will be the same on both sides of the plug and the temperature change may therefore be determined from a set of isenthalps such as those for nitrogen shown in Fig. 1.5(b). The heavy curve indicates the locus of their maxima and is known as the *inversion curve*. For pressures to the right of the inversion curve, the temperature of the gas will increase slightly when it passes through the plug; and for those to the left, there will be a small decrease of temperature. The most effective choice of P_i, for a given value of P_f, will clearly be a point actually on the inversion curve. For temperatures above the *maximum inversion temperature* T_{inv}, defined as the upper of the two points at

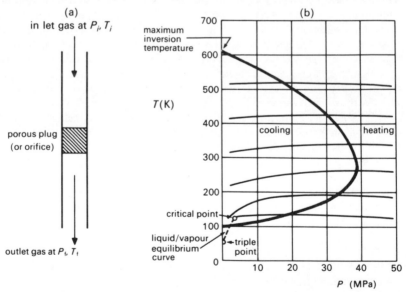

Figure 1.5 The Joule–Kelvin effect. (a) When gas is allowed to expand through a porous plug or orifice it may become either warmer or cooler, but its molar enthalpy remains constant. (b) Examples of isenthalps, for nitrogen. For pressures to the left of the inversion curve (bold line) cooling will occur on expansion.

which the inversion curve intersects the temperature axis, the expansion invariably results in heating. Thus, if the Joule–Kelvin effect is to be employed in a liquefier, it is essential to precool the gas to a temperature well below T_{inv} before expanding it. Some values of T_{inv} for typical cryogens are given in Table 1.2. The efficacy of the cooling effect is clearly determined by the gradient of the relevant isenthalp, which is known as the Joule–Kelvin coefficient,

$$\mu = \left(\frac{\partial T}{\partial P}\right)_h.$$ (1.8)

It is straightforward to demonstrate that

$$\mu = \frac{1}{C_P}\left[T\left(\frac{\partial v}{\partial T}\right)_P - v \right]$$ (1.9)

where v and C_P are respectively the molar volume and molar heat capacity at constant pressure. For an ideal gas, $T(\partial v/\partial T)_P$ is equal to v, so that $\mu = 0$, corresponding to a zero temperature change as the gas passes through the plug. It will be noted that one particular isenthalp in Fig. 1.5(b) ends at the critical point. Isenthalps lying lower than this must intersect the liquid–vapour equilibrium curve (dashed, joining the critical point to the triple point) which implies that at least some of the gas will then liquefy.

The provision of counterflow *heat exchangers* is usually crucial to the design of a liquefier and the optimization of its performance. These commonly consist of a pair of concentric tubes in which the hot incoming and cold outgoing gas streams flow in opposite directions, so that the former is precooled by the latter. A variety of manufacturing techniques (for example, the use of coiled coils or the attachment of helical fins) is used to increase the effective surface area of the exchanger and thereby to promote the flow of heat between the two gas streams.

The use of a heat exchanger makes it possible to liquefy air by Joule–Kelvin expansion alone, starting from room temperature (which is considerably smaller than T_{inv} for either nitrogen or oxygen: see Table 1.2). The operating principle is sketched in Fig. 1.6(a). Upon first starting up, some cooling, but no liquid, is produced as the high pressure incoming air expands at the Joule–

Table 1.2 Maximum inversion temperatures for some selected cryogens

Substance	T_{inv} (K)
Oxygen	893
Nitrogen	607
Hydrogen	204
Helium-4	43

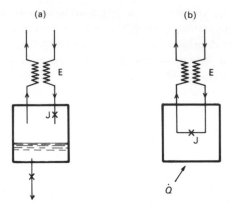

Figure 1.6 (*a*) Principle of operation of liquefier based on Joule–Kelvin expansion. The gas expands through valve *J*, some of it liquefies, and the outgoing cold gas cools the incoming gas stream through the heat exchanger *E*. (*b*) Virtually the same arrangement will serve as a refrigerator in which a heat flow $\dot{Q}$ can be absorbed at the cold end of the system, but no liquid is collected.

Kelvin expansion valve *J*. The resultant colder, low-pressure, air flows back through the heat exchanger *E* and cools the incoming air stream. As a result, the whole system gradually cools. Eventually, steady operating conditions are reached and a certain fraction of the incoming air then liquefies as it expands through *J*. This arrangement is often known as the *Linde–Hampson* system. Figure 1.6(*b*) shows how almost the same configuration can be made to serve as a refrigerator: there is then a cold region where a flow of heat $\dot{Q}$ may be absorbed, but no liquid is collected. It must again be emphasized that these simple schemes would not, of course, work for gases such as hydrogen or helium, for which T_{inv} is considerably below room temperature and which therefore need to be precooled before the expansion.

Another single-stage air liquefaction process uses the Stirling cycle and forms the basis of the popular Philips nitrogen liquefiers. The working fluid within the Stirling engine is usually helium, and is operated on in a closed cycle, with the regenerator effectively performing the functions of a heat exchanger in separating the colder and hotter ends of the system. Air condensation surfaces are attached to the outside of what corresponds to the colder cylinder of Fig. 1.4(*a*). For the production of liquid nitrogen, which is usually to be preferred over liquid air on grounds of safety and constant temperature, a distillation column is provided for removing the unwanted oxygen.

Most liquefiers, however, incorporate several stages of cooling and involve judicious combinations of the three basic techniques described above. The final stage is usually a Joule–Kelvin expansion. The incoming gas is precooled: either by exchange of heat with a bath (or baths) of more readily liquefied cryogen(s), known as the *cascade* process; or by passing some proportion of it through an expansion engine (or engines). The cascade process can be designed, with suitable heat exchangers, to have a high thermodynamic efficiency, but it tends to be rather cumbersome. As an example of an external work/Joule–Kelvin combination liquefier, we sketch in Fig. 1.7 a simplified flow diagram for the A.D. Little/Collins helium liquefier which has been used in many low-temperature laboratories throughout the world since 1946. There are two piston-in-cylinder expansion engines whose respective working temperatures (approximately) are 60 K/30 K and 15 K/9 K. The incoming helium gas enters via a series of heat exchangers and is distributed roughly as follows: hotter engine, 30%; colder engine, 55%; Joule–Kelvin stage, 15%. Very careful design of the mechanical components enables them to operate at temperatures where conventional lubrication methods are impossible. Scrupulous gas purification is obviously essential because even quite small traces of air, for example, can solidify and cause seizure of the piston. Provision for an alternative, and usually better, method of positioning the Joule–Kelvin valve is also provided. It may be placed external to the liquefier (to which it is connected via a triaxial transfer tube, with vacuum surrounding the two counter-flowing gas streams) so that the final expansion to produce the liquid helium occurs within the actual vessel in which the liquid is to be stored.

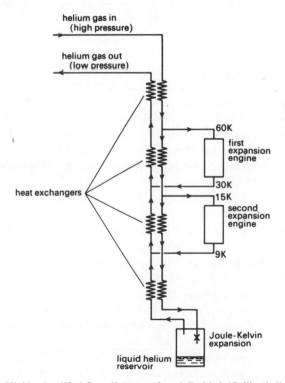

Figure 1.7 Highly simplified flow diagram of an A.D. Little/Collins helium liquefier.

Very much more detailed descriptions of a wide range of different types of liquefier and refrigerator are given in the list of further reading at the end of the chapter, notably in Barron (1966) and Zemansky and Dittman (1981).

1.4 Solids at low temperatures

All matter, except helium, becomes solid when cooled to a low enough temperature. Helium remains liquid at the lowest temperatures which can be attained, although even helium succumbs to solidification on the application of sufficient external pressure (see §1.5). The low temperature solid usually has properties which are very different from those of solids at room temperature and it may, or may not, be crystalline in structure.

At temperatures near 1 K most solids have an extremely small molar heat capacity, C_V, usually amounting to less than a thousandth of the room temperature value. Their thermal conductivities κ cover a very much wider range than is found at room temperature and can be as high as 10^3 W K^{-1}m^{-1} for copper or as low as 10^{-3} W K^{-1}m^{-1} for graphite. In general, however, the coefficient of thermal diffusivity $\kappa/\rho C_V$, where ρ is the density, is considerably larger than the value found at room temperature. As a result, thermal equilibrium times tend to be extremely short: a most convenient feature of measurements at low temperatures. The flow of heat from one body to another occurs less easily at low temperatures, however, corresponding to a relatively high value of the *Kapitza resistance*, a point to which we return below. The coefficient of thermal expansion is negligible. Electrical conductivities of alloys are usually much the same as at room temperature, but those of pure metals can become enormously larger at low temperatures, with an increase over the room temperature value by a factor as much as 10^5 in the cases of extremely pure single crystals of, for example, tin, cadmium or zinc. The physics underlying these interesting changes in properties is discussed in chapters 2 and 3, and their relevance to the design of cryostats and other low temperature apparatus in chapters 7 and 8.

Superconductivity is a phenomenon for which no known analogue exists at room temperature. For certain materials, the electrical resistance vanishes totally at low enough temperatures. If a circulating current is made to flow around a closed loop of the superconductor, it is found to flow indefinitely without any observable decay, although its presence may readily be detected from the magnetic field that is generated. In pure metals, the transition to the superconducting state occurs at a well-defined critical temperature T_c and, as the width of this transition can be extremely narrow (10^{-5} K in pure gallium), the critical temperatures provide valuable fixed points on the thermometer scale.

In addition to possessing the property of perfect conductivity, super-conductors are also perfect diamagnets. A magnetic field is excluded from the bulk of the material irrespective of whether the field was applied before or after cooling below T_c. This is not an effect that would be expected in a normal metal with zero resistance. It implies the existence of a critical magnetic field H_c, because the exclusion of magnetic field from the superconductor increases its free energy: in zero applied field, the superconducting state must have a lower energy than the same metal in the normal state at the same temperature, so that the superconducting state must be destroyed when the field energy equals this condensation energy. Thus the transition to the superconducting state is a function of temperature T and the applied field H which can be represented by the simple phase diagram shown in Fig. 1.8.

Superconductivity is found in a wide range of materials: metallic elements, alloys, intermetallic compounds and, recently, synthetic organic solids, which in the normal state are poor conductors. Although the fundamental properties of perfect conductivity and perfect diamagnetization are common to all superconductors, variations occur in some of their thermal, magnetic and electrodynamic properties. As we will see in chapter 4, however, the microscopic theory of Bardeen, Cooper and Schrieffer provides a general basis for understanding the behaviour of almost all superconductors. At this stage it should be mentioned that superconductors can be divided into two types, depending on their behaviour in an applied magnetic field. The distinction is made on the basis of how flux penetrates into the bulk of the material. In an ideal situation, the flux does not penetrate into a type I superconductor until the critical field is reached, and then it does so discontinuously. By contrast, in a type II superconductor, partial flux penetration can occur over a wide range of applied field values, even though the zero resistance property may be maintained up to a limiting applied field of more than 20T in some materials. This property is of particular importance in the technological exploitation of superconductors for high-current and high-field applications. We discuss the phenomenon of superconductivity in more detail in chapter 4 and we describe some of its more important applications in chapter 8.

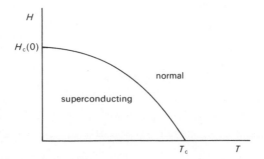

Figure 1.8 The phase diagram for a superconductor. It is found experimentally that

$$H_c(T) = H_c(0)[1 - (T/T_c)^2].$$

1.5 Liquid helium

The two stable isotopes of helium, ^{4}He and ^{3}He, are indispensable to cryogenics. Naturally-occurring helium consists almost exclusively of ^{4}He but it also contains traces ($c.\ 10^{-7}$ in well helium, $c.\ 10^{-6}$ in atmospheric helium) of ^{3}He. Most of the ^{3}He used for cryogenic purposes, however, is now prepared as a byproduct of tritium manufacture and storage, through the neutron bombardment of lithium:

$$^6_3\text{Li} + ^1_0\text{n} \rightarrow ^3_1\text{H} + ^4_2\text{He} \tag{1.10}$$

$$^3_1\text{H} \xrightarrow{12.5\text{yr}} ^3_2\text{He} + \text{e}^- + \bar{\nu} \tag{1.11}$$

The unstable helium isotope ^{6}He has been used in low-temperature experiments, notwithstanding its relatively short half-life of 0.8s, but we shall not consider it further here.

The low-temperature P–T phase diagrams of ^{4}He and of ^{3}He in zero magnetic field are shown in Fig. 1.9(a) and (b) respectively, plotted on the same

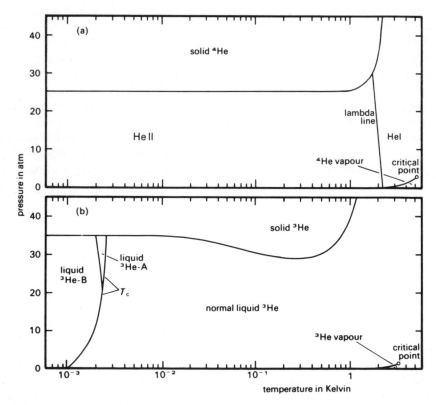

Figure 1.9 Low-temperature phase diagrams (a) for ^{4}He, and (b) for ^{3}He under zero magnetic field.

scale for convenient comparison. Although they share certain features in common, some very striking differences are immediately apparent. Both isotopes form low-density liquids when cooled below a few Kelvins and both would apparently remain liquid right down to absolute zero under their saturated vapour pressures; but both can be solidified by the application of external pressure, with more being required in the case of ^{3}He (a limiting low-temperature value of 34 atm) than in that of ^{4}He (25 atm). The ^{3}He solid/liquid equilibrium line passes through a well-defined minimum near 0.3K, a most interesting feature to which we return in §6.2. The saturated vapour pressure of ^{3}He is considerably larger than that of ^{4}He for any given temperature; and its critical point (Table 1.1) lies below that of ^{4}He.

Both ^{3}He and ^{4}He exhibit a quite extraordinarily rich and diverse range of physical phenomena at very low temperatures, known collectively as *superfluid properties*. These include, for example: a capability for frictionless flow through tiny holes or pores; quantized rotational modes; and an ability to support temperature/entropy waves propagating through the liquid with minimal damping, much like the familiar pressure/density waves of ordinary sound. For both isotopes, the onset of superfluidity occurs with extreme suddenness, as the liquid is cooled, at a transition temperature whose value is dependent on pressure.

For ^{4}He, onset occurs at what is known as the *lambda temperature*, T_λ, so named because the shape of the corresponding specific heat anomaly is reminiscent of a Greek lambda. T_λ has a value of 2.17K under the saturated vapour pressure and decreases with increasing pressure to 1.80K at the solidification point. The change in properties at T_λ is so abrupt and so remarkable that it is as though the experimentalist were dealing with two different liquids, above and below the transition. Accordingly, the liquid phases of ^{4}He above and below T_λ are distinguished by referring to them as HeI and HeII, respectively. The lambda line separating the two phases is plotted in Fig. 1.9(a).

In the case of ^{3}He, the onset of superfluidity occurs at a temperature T_c which is smaller than T_λ by a factor of about a thousand; and there are two (principal) superfluid phases rather than one, known as ^{3}He-A and ^{3}He-B, as indicated in Fig. 1.9(b). T_c takes a value of 1.0 mK under the saturated vapour pressure (which is effectively zero) but increases to 2.8 mK at the solidification point. The non-superfluid phase of liquid ^{3}He is a system of considerable interest in its own right. Usually referred to as *normal liquid ^{3}He*, it possesses quantal properties at low temperatures which make it utterly different from HeI, HeII, ^{3}He-A or ^{3}He-B.

The properties of liquid ^{4}He and liquid ^{3}He are discussed in more detail in chapters 5 and 6 respectively. Here we shall simply consider why helium exists as a liquid at all at very low temperatures. This unique and at first sight astonishing behaviour results from the combined effects of the exceptionally feeble interatomic forces and the very low atomic mass.

The helium atom possesses an inert gas structure, with its two electrons forming a closed K shell. The only attraction between a pair of helium atoms, therefore, is that due to the van der Waals' force, which arises because the fluctuating electric dipole on one atom tends to induce a polarization of the other atom (since we can ignore, for now, the even tinier force due to the interaction between the nuclear magnetic dipole moments in the particular case of a pair of ^{3}He atoms). In fact, the helium atom is a particularly rigid spherical structure and it is not at all easily polarized. Indeed, it has the smallest polarizability of any atom and, consequently, the weakest van der Waals' force. In many ways, it approximates remarkably well to the hard sphere assumed by classical kinetic theory. One might suppose that, despite the weakness of the interatomic forces, the internal energy of crystalline helium would inevitably be less than that of the liquid phase, albeit only slightly less, so that the system would be bound to solidify once the temperature had been sufficiently reduced. Such a view ignores important quantum mechanical considerations, however. In particular, it takes no account of the crucial role played by the zero-point energy (London, 1954).

The quantum mechanical zero-point energy of a confined particle (energy of confinement) increases as the size of its confining box is reduced, and is inversely proportional to the mass of the particle. This may readily be demonstrated by means of a simple argument based on the Heisenberg uncertainty principle:

$$\Delta p \Delta x \geqslant \hbar. \tag{1.12}$$

In the case of liquid helium, we suppose that any given atom is confined to the cavity of average radius R formed by its neighbours, so that the uncertainty of its position $\Delta x \simeq R$. Thus the uncertainty in momentum $\Delta p > \hbar/R$. Consequently, the minimum uncertainty in energy, which corresponds to its zero-point energy,

$$E_Z \simeq (\Delta p)^2/2m \simeq \hbar^2/2mR^2. \tag{1.13}$$

The low mass of the helium atom will therefore imply a large value of E_Z.

London calculated the total energy of both the liquid and solid phases of ^{4}He at zero temperature and pressure, with the results sketched in Fig. 1.10. His estimate of the potential energies arising from the interatomic force are shown as a function of molar volume by the lower two curves of Fig. 1.10(a). Not surprisingly, the solid has the deeper energy minimum and, in the absence of other considerations, it would clearly be the stable phase. When the contribution of the zero-point energy E_Z (upper curve) is added in, however, the situation is transformed. The estimated total energies of the two phases are then as shown in Fig. 1.10(b), where the liquid clearly has a lower minimum energy than the solid. The stable phase at $T = 0$, $P = 0$ must therefore be the liquid. It is also evident from a comparison of (a) and (b) of Fig. 1.10 that the molar volume is a great deal larger than would be expected for the classical

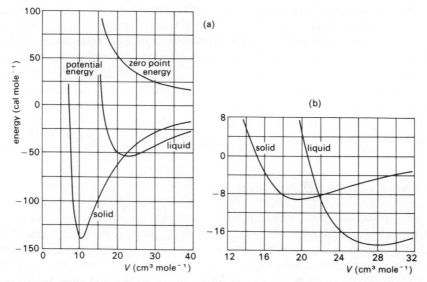

Figure 1.10 Illustration (after F. London, 1954) of how zero-point energy prevents ^{4}He from solidifying under its saturated vapour pressure. Note the change of scales between (a) and (b) (and that 1 calorie = 4.2 joules).

assembly of hard spheres, whence London's evocative description of helium as 'a quantum liquid blown up by its own zero-point energy'.

Although the above discussion relates to liquid ^{4}He, the arguments are applicable *a fortiori* to liquid ^{3}He, which possesses interatomic forces almost identical in magnitude to those of liquid ^{4}He, but which has an even smaller atomic mass. Zero-point energy raises the low temperature molar volume of liquid ^{4}He to about 3 times the volume that would otherwise be expected; and that of ^{3}He to about 4 times the expected value. It should be noted that the application of similar arguments to other substances invariably leads to the conclusion that they will be solid at $T = 0$. This is because the interatomic or intermolecular forces are much stronger than for helium and because, except in the case of hydrogen, the mass is larger, thereby reducing the zero-point energy. Consequently, the potential energy minima of both liquid and solid will be considerably deeper than those of Fig. 1.10 and the relative contribution made by the zero-point energy to the total energy will be a great deal smaller. In the particular case of hydrogen, the strong intermolecular forces overcome the effect of the relatively large zero-point energy, thereby making hydrogen a solid at low temperatures.

The large molar volume of liquid helium gives it a very loose structure compared to other liquids. One consequence is that the viscosity of even the non-superfluid phases near their normal boiling points is considerably less than that of air at room temperature. We shall see in chapters 5 and 6 that the key to an understanding of the liquid heliums lies in the realization that, for many purposes, it is much better to regard them not as liquids at all, but as rather non-ideal gases. This approach makes it possible to account for the enormous differences in many of the properties of liquid ^{3}He and liquid ^{4}He, because their behaviour will be governed by different sets of quantum statistics (Bose–Einstein statistics for ^{4}He and Fermi–Dirac statistics for ^{3}He). It also helps to illuminate the seemingly vexatious question of why the existence of liquids at $T = 0$ does not, in itself, constitute a violation of the Third Law: both liquids do, in fact, become highly ordered at low temperatures, but the ordering takes place in momentum space rather than in ordinary Cartesian space.

1.6 Macroscopic quantization

Superfluidity is a property which HeII, ^{3}He-A and -B, and the electron gas in superconducting metals, all have in common. The principal experimental observation in each case is that fluid flow can be maintained without dissipation and in the absence of any applied driving force. In the cases of HeII and superconductors (at least) the flow can take the form of a persistent current in a closed loop, continuing indefinitely without diminution. Similarly, it is found that an object can move through stationary HeII without experiencing any drag force, provided that the temperature is low enough; and it seems very probable that the same result will be obtained for ^{3}He when the appropriate measurements have been extended to sufficiently low temperatures. For the electron gas in a superconductor, however, the presence of the lattice clearly prohibits the latter type of experiment.

These are three very different systems. The ^{4}He atom is electrically neutral and magnetically inert; the electron is charged and carries a magnetic moment; the ^{3}He atom is uncharged but has a tiny nuclear dipole moment; and, in addition, the helium masses are $c.$ 6000 times larger than that of an electron. Nonetheless, despite these huge differences, it is only natural to seek a common physical mechanism, responsible for the appearance of superfluidity in all three cases.

We shall see in chapters 4–6 that there is strong evidence to suggest that the crucial feature which the three systems have in common is that they each possess a *condensate*, implying the occupation of a single quantum state by a macroscopic number of particles. By analogy with the ideal Bose–Einstein gas, macroscopic numbers of ^{4}He atoms may be expected to condense (see §5.1) into the zero momentum ground state below a certain characteristic temperature. A similar process for ^{3}He atoms or for electrons seems, at first sight, to be impossible because they are fermions (for which only one particle per quantum state is permitted). We shall see, however, that an assembly of electrons or of ^{3}He atoms can sometimes reduce its total energy by forming *pairs* of particles. The particle pairs, being bosons, can then undergo Bose–Einstein condensation, much like ^{4}He.

London (1954) argued that it was the particles in the condensate which formed the supercurrent in each case (something of an over-simplification, as it turns out, but the general line of argument is correct). The only other known currents able to flow without any dissipation were the orbiting electrons in atoms and molecules, each of which occupies a stationary quantum state which is an eigenfunction of the relevant Hamiltonian. By analogy, therefore, supercurrents might also be quantum currents, describable in terms of a *macroscopic wave function* which extended throughout the entire sample. Following London, therefore, we postulate that there exists a wavefunction ψ which, under steady-state conditions, may be written in the form

$$\psi(\mathbf{r}) = \psi_0 e^{iS(\mathbf{r})} \qquad (1.14)$$

where the phase $S(\mathbf{r})$ is a real function of position $\mathbf{r}$. We suppose, for now, that the amplitude ψ_0 is a scalar quantity. We further assume that ψ will be governed by the ordinary single-particle wave equation. Thus, for zero potential-energy gradient,

$$-i\hbar\nabla\psi = \mathbf{p}\psi \qquad (1.15)$$

whence, substituting for ψ

$$\mathbf{p} = \hbar\nabla S. \qquad (1.16)$$

Considering, first, the application of these ideas to HeII at zero temperature, we may interpret (1.16) in terms of the motion of one atom of superfluid, so that

$$\mathbf{p} = m_4 \mathbf{v}_s \qquad (1.17)$$

where $\mathbf{v}_s$ is the superfluid velocity. Then

$$\mathbf{v}_s = \frac{\hbar}{m_4}\nabla S \qquad (1.18)$$

or, in other words, the velocity of superflow is proportional to the gradient of the phase of the wavefunction (1.14). When the superfluid is stationary, the phase is uniform throughout; but otherwise, the local value of the phase S varies with position at a rate proportional to $\mathbf{v}_s$, along the direction of flow. Substitutions of numerical values demonstrate immediately that, for typical values of $\mathbf{v}_s$ used in experiments, S varies extremely slowly on an atomic scale: uniform flow at $10\,\mathrm{mm\,s^{-1}}$, for example, would imply a wavelength for ψ equivalent to $c.\ 3 \times 10^4$ average interatomic spacings. This property of *phase coherence* over macroscopic regions of the liquid implies a correlation of the individual motions of huge numbers of atoms. It provides an immediate, though qualitative, understanding of the origins of superfluidity, because any alteration in the motion of one particular atom must require a corresponding simultaneous modification to the motions of a macroscopic number of other atoms, which clearly constitutes a relatively improbable event.

Secondly, much the same sort of argument can be applied to the superfluid phases of liquid ^{3}He (which were, of course, still undiscovered when London proposed his original ideas). An important difference, however, to which we will return in §6.4, is that the 'particles' of the system are now pairs of atoms, of mass $2m_3$, so that (1.18) is replaced by

$$\mathbf{v}_s = \frac{\hbar}{2m_3}\nabla S \qquad (1.19)$$

The comments made above on the nature of superfluidity are still applicable although, as will become clear in §6.4, ψ_0 in (1.14) must be replaced by a vector quantity.

Thirdly, in applying similar arguments to the electron gas in a super-conductor, we will assume that the 'particles' have a mass m and a charge $-q$. In chapter 4, we will see that these particles are in fact pairs of electrons; but, irrespective of the exact nature of the particles, it is essential to take account of the fact that, for a charged particle moving in a magnetic field of vector potential $\mathbf{A}$, the relationship between velocity and momentum will be given by

$$\mathbf{p} = m\mathbf{v}_s + q\mathbf{A} \qquad (1.20)$$

(rather than by the simpler form (1.17), applicable to a neutral particle). Eliminating $\mathbf{p}$ between (1.16) and (1.20) we obtain for the electron superfluid velocity

$$\mathbf{v}_s = \frac{\hbar}{m}\nabla S - \frac{q\mathbf{A}}{m} \qquad (1.21)$$

Again, the comments about phase coherence and superfluidity in relation to ^{4}He are equally applicable to superconductors. Indeed, they apply with redoubled force because, for any given flow velocity, the smaller mass of the 'particles' in a superconductor implies, through (1.21), an even more gradual change in ψ with position than is the case for the liquid helium. Where two pieces of superconductor are weakly coupled together, for example by a very thin oxide barrier, the mutual interaction of their wavefunctions gives rise to a wide variety of interesting and technologically important phenomena known collectively as the *Josephson effects*. These are discussed in §4.6.

Although London's postulate of a macroscopic wavefunction represented a very considerable leap of faith at the time he proposed it, there is now a substantial body of evidence that the general framework of his ideas is indeed correct. We refer to them in more detail in relation to particular experimental observations in chapters 4–6. A particularly full and stimulating discussion of the macroscopic wavefunction, and its associated phenomena, will be found in the book by Tilley and Tilley (1974).

There is a continuing debate about the extent to which the phenomena of macroscopic quantization and superfluidity represent universal behaviour, to be expected of any assembly of freely mobile particles at low enough temperatures ('low' here, being in the second of the senses discussed in §1.1, implying that the temperature/density quotient is small). It is possible, for example, that, at sufficiently low temperatures, *all* non-magnetic metals may become superconductors, not just the particular metals (see chapter 4) now known to do so. It is also believed that the interiors of certain types of stars may be superfluid (see §6.6). Indeed, superfluidity may be a great deal more common in the universe than we might at first suppose, given the difficulty, until quite recently, of demonstrating the phenomenon in the laboratory. Mendelssohn (1977) develops this line of speculation even further:

> While on earth the superfluids are hidden at the lowest temperatures, the study of this obscure region may have revealed to us the most prevalent state of condensed matter in the universe. In fact, when compared with superfluidity, the solid state of crystalline aggregation may turn out to be a rare and quaint exception confined to highly unrepresentative conditions, such as the surface of this planet.

Bibliography

Further reading

Barron, R. *Cryogenic Systems*. McGraw-Hill, New York and London (1966).

Dugdale, J.S. *Entropy and Low Temperature Physics*. Hutchinson, London (1966).

London, F. *Superfluids*. Vol. I, *Superconductivity*, Vol. II, *Superfluid Helium* (1954); reprinted, Dover, New York (1964).

MacDonald, D.K.C. *Introductory Statistical Mechanics for Physicists*. Wiley, New York and London (1963).

Mendelssohn, K. *The Quest for Absolute Zero: the Meaning of Low Temperature Physics*, 2nd edn,. Taylor and Francis, London (1977). (An historical perspective, readily accessible even to non-specialists.)

Pippard, A.B. *Elements of Classical Thermodynamics*. Cambridge University Press (1957).

Rosenberg, H.M. *Low Temperature Solid State Physics*. Clarendon Press, Oxford (1963).

Simon, F.E. In *Low Temperature Physics–Four Lectures*. Pergamon, London (1952). (A general review, containing some ideas that are still of topical interest, even after a lapse of more than thirty years.)

Tilley, D.R. and Tilley, J. *Superfluidity and Superconductivity*. Van Nostrand Reinhold, London (1974); and new edition (Adam Hilger, forthcoming). (Emphasizes, particularly, the role played by the macroscopic wave function.)

Wilks, J. *The Third Law of Thermodynamics*. Oxford University Press (1961).

Zemansky, M.W. and Dittman, R.H. *Heat and Thermodynamics*. 6th edn., McGraw-Hill, New York (1981).

2. Lattice vibrations

2.1 The model of a solid at low temperatures

Solids usually occur as polycrystalline aggregates. Although their macroscopic physical properties may then be the average of a large number of randomly oriented microcrystals, nevertheless the local environment of almost every individual atom is totally ordered. Thus attempts to understand the physical properties of most solids begin from the model of a single crystal, with the atoms arranged on an infinite lattice. In contrast, a small number of materials occur in an amorphous state with the disordered structure of a glass. Such a structure is not thermodynamically stable, however, and can be visualized as a supercooled liquid that is slowly relaxing back to its equilibrium crystalline state over a long period of time. It is only recently that significant advances have been made in our understanding of amorphous structures as the result, particularly, of low-temperature experiments. We shall review some of the current ideas in §2.7 and §3.5, but the main theme of chapters 2 and 3 refers to crystalline solids. In the present chapter, we shall describe some of the low-temperature phenomena that arise from the motion of the atoms in the crystal lattice, that is, the lattice vibrations, or phonons. In the following chapter, we shall discuss the effects that are due to the electrons. In studying solids, great simplification can often be achieved by making the measurements at as low a temperature as possible. In general terms, the reason for this is that as the temperature approaches more closely to absolute zero, so do the properties of real solids agree more exactly with those of the ideal models on which their description is based. A good illustration of this principle is with reference to the *harmonic approximation* and the effects that follow from it. Since the idea is also fundamental to a description of lattice vibrations, we spend a few moments examining the details of the harmonic approximation. (A more rigorous development will be found in, for example, Ashcroft and Mermin, 1976.) Each atom or ion in a crystal is confined in a potential 'well' because of the electrostatic forces of its neighbours, as shown in Fig. 2.1. The exact shape of the well, that is to say, the dependence of potential energy, V, on the atomic separation, r, cannot be calculated analytically except for a very few

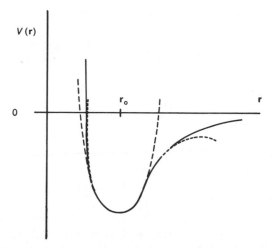

Figure 2.1   The potential 'well' occupied by an atom in a crystal, illustrating the modelling of the potential energy $V(\mathbf{r})$ by a Taylor series in atomic separation $\mathbf{r}$.
– – – – harmonic approximation
.... with a cubic anharmonic term
———— actual potential

structures. More commonly a Taylor expansion of V in terms of $\mathbf{r}$ is used and the potential energy is usually written

$$V(\mathbf{r} - \mathbf{r}_0) = C(\mathbf{r} - \mathbf{r}_0)^2 - D(\mathbf{r} - \mathbf{r}_0)^3 - E(\mathbf{r} - \mathbf{r}_0)^4 + \cdots \qquad (2.1)$$

The equilibrium separation of the atoms is $\mathbf{r}_0$, in the absence of anything to disturb them such as lattice vibrations. There can be no linear term $B(\mathbf{r} - \mathbf{r}_0)$ since this would imply a constant force acting on each atom, so that $\mathbf{r}_0$ could not, after all, be the equilibrium coordinate. Of the coefficients C, D, and E, the first is by far the largest in most materials.

The harmonic approximation consists of setting D and E explicitly equal to zero in the expansion above. Thus the shape of the potential well is defined as parabolic, the restoring force acting on an atom is assumed to be directly proportional to its displacement from equilibrium, and the whole model is simplified to that of the familiar harmonic oscillator. The description of lattice vibrations and the phenomena which arise from them takes off from those simple assumptions, as will be described in §2.2. When lattice vibrations are excited thermally, it is at low temperatures that the atoms stray least from the minima of their potential wells and the harmonic approximation therefore has the greatest validity. At higher temperatures, the increasing effect of the cubic, D, and quartic, E, terms in (2.1) gives rise to various anharmonic phenomena. Thermal expansion is probably the most familiar of these, but equally important in solid-state physics is phonon–phonon scattering, described in §2.5. In the so-called quantum crystals, particularly solid helium, the anharmonic effects are very large indeed, giving rise to new phenomena which we shall describe in §2.6.

Another approximation employed in modelling the low temperature properties of solids involves the concept of an elementary excitation. Often it is simpler to visualize the behaviour of a crystal in terms not of the absolute coordinates of the individual ions and electrons, but of the deviations of those coordinates from a set of ground-state values. Indeed, in many situations it is the only possible analytical approach to the problem. Expressed another way, we say that most physical properties of solids at low temperatures are determined by the low-lying excited states of the crystal, rather than by the actual binding energy of the ground state itself. For example, lattice vibrations, expressed as deviations from $r = r_0$, are elementary excitations. Similarly, as discussed below in chapters 3 and 4, many of the properties of metals and semiconductors are determined by low-lying excited states that depend on electrons and holes. Other examples of elementary excitations include rotons and phonons in liquid helium (chapter 5) and quasiparticles in super-conductors (chapter 4).

Elementary excitations, then, correspond to small deviations of a physical system from its ground state. The energies of the excitations are defined relative to the ground-state energy. The number of excitations actually present in the crystal at a particular temperature is determined by the appropriate distribution function, either Fermi–Dirac or Bose–Einstein depending on the spin of the excitation. Phonons and photons are bosons, whilst electrons are fermions. In any particular solid at a given temperature there may be contributions to the free energy, and hence to its physical properties, from several different types of elementary excitations. In metals, for example, there are significant contributions to the free energy from both the electrons and phonons. At temperatures below 1K the electronic heat capacity is larger, whilst at higher temperatures the phonon contribution dominates. It is often possible in considering the combined effect of different types of elementary excitations to make the assumption that they can be regarded as totally independent, so that the magnitude of a particular physical property of the solid can be taken simply as the sum of the contributions of the individual excitations. As might be expected, this assumption is never completely valid in real materials, but it is certainly closer to the truth at low temperatures since fewer excitations of any type are present. However, many important pheno-mena actually result from a modest interaction of one excitation with another. Thus thermal and electrical resistivities are determined largely by the interactions of phonons and electrons giving rise to scattering. It is worth noting, also, that in some situations the interaction between two types of excitation can be so large that it is impossible to use the approximation of independence even as a starting point. In ionic crystals like sodium chloride for instance, electrons and phonons are so strongly coupled that a totally new excitation emerges, called a polaron, possessing components both of electric field and of lattice distortion. But the most striking manifestation of the electron–phonon interaction is the formation of a 'Cooper pair' of electrons, giving rise to the phenomenon of superconductivity. We begin with phonons.

2.2 Lattice vibrations—basic ideas

The formal theory of lattice vibrations has been described in many textbooks, such as Ashcroft and Mermin (1976) and Kittel (1976). In the present section we shall briefly review fundamental ideas in order to be able to concentrate on those aspects which are of special interest at low temperatures.

Much of the basic physics of lattice vibrations can be understood from a consideration of the very simplest model, that of a one-dimensional lattice with one atom per lattice point. In the harmonic approximation, as we saw, each atom is assumed to be confined in a parabolic potential well

$$V(x) = \frac{F}{2}(x - x_0)^2 \tag{2.2}$$

The equilibrium position, x_0, for an atom is not fixed with respect to the crystal as a whole since it depends on the relative positions of the neighbours, and they themselves are also in motion. The analogy of a line of heavy balls, each of mass m, connected together by harmonic springs, each of length a, and restoring force constant F, is a helpful one (Fig. 2.2). If u_n is the displacement of the nth atom, then the equation of motion can be written

$$m\frac{d^2 u_n}{dt^2} = F(u_{n+1} + u_{n-1} - 2u_n) \tag{2.3}$$

Intuition applied to the expected behaviour of the ball-and-spring model suggests a wave-like solution,

$$u_n = u_0 \exp i(\omega t + q x_n) \tag{2.4}$$

Substitution into (2.3) verifies that (2.4) is indeed a valid solution, as long as

$$\omega = \pm 2\left(\frac{F}{m}\right)^{1/2} \sin \frac{qa}{2} \tag{2.5}$$

The first point to be noted about this *dispersion relation*, plotted out in Fig. 2.2, is that it is periodic in the wavevector, q. Any wave, regardless of its 'true' wavevector, has the same frequency as a 'reduced' wavevector somewhere in the first Brillouin zone, between the limits $\pm \pi/a$. The two values of q are related by the simple addition or subtraction of multiples of $2\pi/a$. Since ω is not a linear function of q, the group velocity of the lattice wave, $d\omega/dq$, which is the velocity with which the waves can transfer energy, is not equal to the phase velocity ω/q. Indeed, the group velocity goes to zero for phonons at the zone boundary, and for even larger wavevectors becomes negative. These features

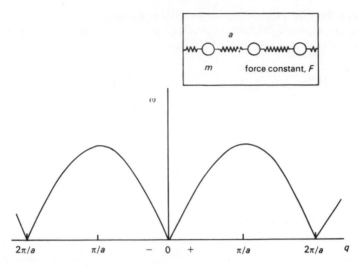

Figure 2.2   The dispersion curve of a ball-and-spring model of a one-dimensional monatomic lattice, as shown in the inset.

will be of particular importance in the later discussion of phonon–phonon scattering.

At low frequencies, the wavelength of the vibrations is long enough not be affected significantly by the discreteness of the lattice, and the group velocity and phase velocity both become closely equal to $(Fa^2/m)^{\frac{1}{2}}$. This *continuum approximation* is widely applicable to low-temperature phenomena, since it is then the low-frequency modes of the lattice that are predominantly excited. For example, the thermal distribution of phonons for a temperature of 4.2 K, the boiling point of ^{4}He, has its maximum value at a frequency of 2.5 $\times 10^{11}$ Hz. This is at least a factor of ten lower than the Brillouin zone boundary for most solids, well down the linear portion of the phonon dispersion curve. The Debye theory of the heat capacity due to lattice vibrations, which will be described shortly, is built firmly on the foundations of the continuum approximation, and for most materials gives excellent agreement with experiment.

Up to this point, we have used purely classical ideas to work out the possible modes of vibration which may or may not be excited in particular experimental circumstances. If the vibrations are excited as heat, then quantum mechanics is needed in order to work out the extent to which each mode actually contributes to the thermal energy of the crystal. Fortunately, the harmonic oscillator is one of the great analytical successes of quantum mechanics; the reader is referred to the textbook by Leighton (1959) for an articulate account. It should be noted that the oscillator here is a model, not of the behaviour of any individual atom, but of a normal mode of the whole lattice, involving a weighted superposition of all the individual atomic displacements. We are interested in the total energy of the oscillator, and it is well known that the nth eigenvalue of an oscillator of frequency ω, is given by

$$E_n = (n + \tfrac{1}{2})\hbar\omega \tag{2.6}$$

Energy can thus be exchanged with an oscillator only in quanta of $\hbar\omega$. By analogy with electromagnetic radiation, these quanta of lattice vibrational energy are called *phonons*. They have all the properties of localization and propagation possessed by photons, except that they are confined to the bounds of the crystal. Like photons, they have integral spin (equal to zero, as compared to unity for photons) and can be created or annihilated. Hence they obey the Bose–Einstein distribution function

$$n(\omega) = \frac{1}{e^{\hbar\omega/k_B T} - 1} \tag{2.7}$$

where $n(\omega)$ is the average number of phonons associated with the normal mode of frequency, ω, at temperature T. Even at low temperatures, $n(\omega)$ is usually much larger than $\tfrac{1}{2}$, so that the thermal energy dominates the zero-point energy of the oscillator. But for He the magnitude of the zero-point motion is so large that it cannot be neglected even as a first approximation. Solid ^{3}He and ^{4}He are often called *quantum solids* for this reason.

All these features of one-dimensional monatomic lattice vibrations are retained, and some significant new ones added, when the model is broadened to encompass a real three-dimensional solid with several atoms per unit cell. Firstly, instead of a single longitudinal mode of vibration for each wavevector, $\mathbf{q}$, now a three-dimensional vector, there exist three modes with orthogonal polarizations. The simplifying assumption is often made that, of the three polarizations, one is longitudinal with the atomic displacement parallel to $\mathbf{q}$, and two are transverse, with the displacements normal both to $\mathbf{q}$ and to each other. The velocity of the transverse modes is usually of the order of a half the longitudinal velocity. However, it should be remembered that, in reality, such a situation exists only in perfectly isotropic crystal structures, which are rare, or along special directions, called pure mode axes, in anisotropic structures. In general, the three polarizations, whilst always remaining mutually orthogonal, are all admixtures of the pure modes. Furthermore, the energy transmitted by the lattice vibrations does not necessarily flow in the direction defined by the wavevector. Musgrave (1970), amongst others, explains these subtleties. The consequences of crystal anisotropy show up most clearly in acoustic experiments, for which modes of well-defined frequency, wavevector and polarization can be easily distinguished. In the calculation of thermal properties, averages taken over direction and polarization are usually assumed although, in recent years, it has become clear that the propagation of thermal phonons also displays strong anisotropy at low temperatures. A description of the phenomenon and implications of so-called *phonon focusing* will be given in §2.4.

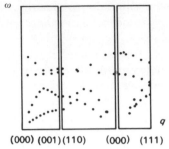

Figure 2.3 Phonon dispersion curves of KBr at 90 K determined by inelastic neutron scattering. The points (000) etc. refer to the reciprocal lattice directions of the phonons. (After Woods *et al.*, 1963.)

The effect of introducing further atoms into the unit cell is to create additional normal modes of the crystal structure. The *acoustic modes*, just described, remain, and correspond to a motion in which all atoms in the unit cell move in the same direction in phase with each other. In addition, however, other modes are possible in which atoms in the unit cell move in different directions although still in phase (or antiphase) with each other. The forces resisting such motion are considerably stronger than those responsible for the acoustic modes, so that the frequencies are higher; they are called the *optic modes* of vibration. If the basis contains p atoms, then there will be $3(p-1)$ optic modes of vibration and 3 acoustic modes. Most solid-state physics textbooks (see, for example, p. 122 of Kittel, 1976) derive the dispersion relation for the lattice vibrations of a one-dimensional diatomic lattice. The dispersion curves for real, three-dimensional solids differ only in detail and can be determined by the technique of inelastic neutron scattering (Egelstaff, 1965). The triple axis spectrometer allows both the energy and wavevector of the incident and scattered neutron to be measured, thus fixing ω and $\mathbf{q}$ for the phonon involved in the scattering. Figure 2.3 shows a dispersion curve of potassium bromide, which is typical of many solids. It is seen that an energy gap of forbidden frequencies exists between the highest acoustic modes and the lowest optic ones, although this is not true for all solids. A second important features of the optic modes is that their group velocity is much lower than that of most acoustic phonons. Because of their higher energies and lower group velocity, optic modes play little part in most low-temperature solid-state phenomena. They are, however, of great relevance to the study of structural phase transitions. In a displacive transition, one of the optic modes in the high-temperature structure softens into a permanent distortion at the transition temperature.

2.3 Heat capacity due to lattice vibrations

The heat capacity is an important quantity in low-temperature physics. In practical terms its magnitude determines the ease with which a particular material may be cooled, whilst in terms of elucidating the physics of a solid, the heat capacity is important as a probe of its energy level distribution. Since the internal energy, U, is impossible to measure directly, a useful second best is a determination of $\partial U/\partial T$ as a function of temperature. In addition, as was seen in chapter 1, the entropy difference ΔS of a system between two temperatures, T_1 and T_2, may be calculated from the temperature variation of the heat capacity at constant volume, C_V. Usually the heat capacity is measured at constant pressure, C_P, so that a correction must be made to obtain C_V:

$$C_P - C_V = 9\alpha^2 \eta T \tag{2.8}$$

where α is the linear coefficient of thermal expansion, and η is the bulk modulus. In practice, the difference is small, especially at low temperatures where thermal expansion becomes disappearingly small.

The heat capacity of diamond as a function of temperature, which is also typical of many other materials, is shown in Fig. 2.4. The explanation of such a curve was one of the first triumphs of the quantum theory, and as such is now part of the folklore of physics, recounted entertainingly by Mendelssohn (1977). In all normal solids, the major contribution to C_V is from the acoustic branch of the lattice vibrations. Even in metals, phonons dominate at all temperatures except those below a few K. The classical prediction for C_V due to the lattice vibrations is easy to obtain. Each atom is assumed to behave as an isolated harmonic oscillator so that the total energy of each atom is

$$E = \frac{m}{2}(v^2 + \omega^2 x^2) \tag{2.9}$$

Mass, velocity, frequency and displacement are represented by m, v, ω and x respectively. Averaged over a Boltzmann distribution, the mean energy per atom is given by

$$\bar{E} = \frac{\int_0^\infty \int \int E \exp\left(\frac{-E}{k_B T}\right) dx \, dv}{\int_0^\infty \int \int \exp\left(\frac{-E}{k_B T}\right) dx \, dv} \tag{2.10}$$

For a crystal which contains N atoms, each able to oscillate in three independent dimensions, $U = 3N\bar{E}$ and C_V is given by

$$C_V = \left(\frac{\partial U}{\partial T}\right)_V = 3 N k_B = 3 \, R \, \text{J} \, \text{K}^{-1} \, \text{mole}^{-1} \tag{2.11}$$

At room temperature this result, known as the law of Dulong and Petit, explains very well both the magnitude and the lack of temperature-dependence of the observed heat capacity for most materials. At lower temperatures, however, the measurements fall increasingly below the predicted value and tend towards zero at the lowest temperatures, as can be seen in Fig. 2.4. It was Einstein, in the first of his two historic papers of 1907, who

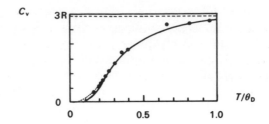

Figure 2.4 The original data on the low-temperature heat capacity of diamond used by Einstein (1907) to support his quantum theory of the heat capacity of solids. The data are compared with
– – – – classical theory
——— Einstein theory, and
·············· Debye theory

pointed out that if energy really were quantized, so that the integrations of 2.11 could be replaced by summations, then much better agreement would be obtained. In this event, equation 2.11 becomes

$$\bar{E} = \frac{\displaystyle\sum_{n=0}^{\infty} E_n \exp\left(\frac{-E_n}{k_B T}\right)}{\displaystyle\sum_{n=0}^{\infty} \exp\left(\frac{-E_n}{k_B T}\right)} \qquad (2.12)$$

The energy E_n, of the nth state of the harmonic oscillator is given by equation 2.6, and the expression is obtained

$$\bar{E} = \left\{ \frac{1}{e^{\hbar\omega_E/k_B T} - 1} + \frac{1}{2} \right\} \hbar\omega_E \qquad (2.13)$$

so that the specific heat becomes

$$C_V = 3Nk_B \frac{\left(\dfrac{\hbar\omega_E}{k_B T}\right)^2 \exp\left(\dfrac{\hbar\omega_E}{k_B T}\right)}{\{e^{\hbar\omega_E/k_B T} - 1\}^2} \qquad (2.14)$$

The quantity ω_E is called the Einstein frequency, and has to be regarded as an adjustable parameter to be fitted individually for each solid. Although the Einstein model gave a greatly improved fit to heat-capacity data, and hence a considerable impetus to the acceptance of the quantum theory, it nevertheless presents a conceptual difficulty. We know very well that the atoms in a solid do not all vibrate independently at the same frequency, but that the crystal lattice as a whole possesses well-defined normal modes, which may be thermally excited. Indeed, as experimental technique improved, it was found that at the lowest temperature a T^3 dependence was observed which could not be explained by Einstein's model which from (2.15) clearly predicts a dependence on $\exp(-\hbar\omega_E/k_B T)$.

The Debye theory rectified this inadequacy, and is still the accepted model of lattice heat capacity at low temperatures. Debye began, as did Einstein, by assuming that a crystal having N lattice points (not the same thing as having N atoms unless the structure is monatomic) could be excited in $3N$ acoustic vibrational modes. The factor of three here refers to the three polarizations associated with each wavevector. The optic modes were neglected as being too high in frequency to be thermally excited at low temperatures. In addition, the continuum approximation was assumed, so that the group velocity and the phase velocity could both be taken as equal to the velocity of elastic waves in the medium, v.

Instead, of an assembly of $3N$ oscillators all of frequency ω_E, in reality there exists a distribution of normal modes of which the number with frequencies between ω and $\omega + d\omega$ is given by $D(\omega)d\omega$, where

$$\int_0^{\omega_D} D(\omega)d\omega = 3N \tag{2.15}$$

and ω_D, the Debye frequency, is the maximum frequency that can be excited. In order to derive an expression for $D(\omega)$, known as the *phonon density of states*, we note that the vibrational modes of the lattice in three dimensions are described by standing waves of the form

$$u(q_x, q_y, q_z) = u_o \exp(i\omega t) \sin q_x x \sin q_y y \sin q_z z \tag{2.16}$$

The derivation of an expression for $D(\omega)$ for such states is given in most textbooks (see Kittel, 1976, page 134), with the result

$$D(\omega) = \frac{3 V\omega^2}{2\pi^2 v^3} \tag{2.17}$$

where V is the sample volume.

We are now in a position to write down an expression for the internal energy, U, of the lattice vibrations, remembering that each mode will be excited to an extent determined by the Bose–Einstein distribution, (2.7). Therefore

$$U = \int_0^{\omega_D} D(\omega)n(\omega)\hbar\omega d\omega$$

$$= \frac{3Vh}{2\pi^2 v^3} \int_0^{\omega_D} \frac{\omega^3}{e^{\hbar\omega/k_B T} - 1} d\omega \tag{2.18}$$

This is most readily evaluated by changing to a dimensionless variable $x = \hbar\omega/k_B T$, and a characteristic temperature, $\theta_D = \hbar\omega_D/k_B$. Then C_V becomes

$$C_V = 9Nk_B \left(\frac{T}{\theta_D}\right)^3 \int_0^{\theta_D/T} \frac{x^4 e^x}{(e^x - 1)^2} dx \tag{2.19}$$

C

In general, analytical methods can take us no further, but numerical solutions of the integral of (2.19) are widely available, for example, in the *American Institute of Physics Handbook* (1963). Our particular interest, however, is in the low-temperature regime where $T \ll \theta_D$. It is then reasonable to take the limits of integration as zero and infinity, whence the value $\pi^4/15$ is obtained for the definite integral. Then

$$C_V = \left(\frac{12\pi^4 N k_B}{5}\right)\left(\frac{T}{\theta_D}\right)^3 \text{J K}^{-1}\text{mole}^{-1} \qquad (2.20)$$

This is the well-known Debye T^3 law, which predicts that the lattice heat capacity is a universal function scaling for all solids through the parameter, θ_D, which is known as the Debye temperature. It is given by the expression

$$\theta_D = \frac{\hbar}{k_B}\left(6\pi^2 v^3 \frac{N}{V}\right)^{1/3} \qquad (2.21)$$

In this expression the quantity v is actually an average over polarization and mode direction, so that C_V is determined largely by the transverse modes, which have the lower velocities. Similarly, materials with strong interatomic forces and light atoms such as as diamond and sapphire have relatively high θ_D, whereas soft materials with low acoustic velocities have smaller values.

Clearly, the path to the Debye T^3 law is strewn with approximations, most of which, following the principle enunciated in §2.1, achieve greater validity

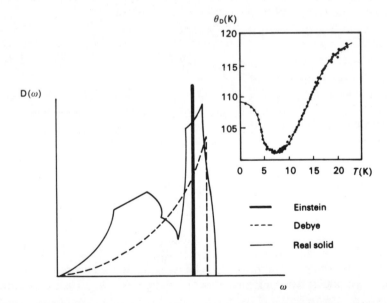

Figure 2.5 The phonon density of states $D(\omega)$ used in the Einstein and Debye theories, compared with that for a typical real solid. The inset shows the variation in the Debye temperature, θ_D, of indium, due to the influence of the extra modes. (After Clement and Quinnell, 1953.)

as the temperature is lowered. The most serious error at slightly higher temperatures is due to the gross oversimplification of the density of states $D(\omega)$. Figure 2.5 illustrates the point by showing schematically the different approximations used for $D(\omega)$, compared with a typical curve measured for a real crystal by inelastic neutron scattering. Nevertheless, the theory provides an excellent fit to low-temperature heat-capacity data. Such slight deviations from T^3 that do exist are conventionally represented as a variation of θ_D with temperature. Figure 2.5 (inset) shows measurements for indium. It is clearly possible to obtain information on $D(\omega)$ from such data but the method has much less resolution than inelastic neutron scattering.

2.4 Thermal conductivity of non-metals

We have seen that heat capacity is an equilibrium property of solids which can normally be understood in terms of the thermal excitation of non-interacting, harmonic, lattice vibrations. In contrast, if heat flow is taking place in a solid, then by definition the system cannot be in thermal equilibrium. There is no universal curve for thermal conductivity, analogous to the Debye theory of specific heat, that can be scaled to fit all solids. Both phonons and electrons contribute significantly to thermal conductivity, although for any but the lowest concentrations of electrons, it is the latter contribution that dominates. In the present section, we shall discuss only the thermal conductivity of non-metals, but even within this limitation there is wide variety. A summary of experimental techniques is given by Berman (1976).

Figure 2.6 shows schematically the temperature dependence of the thermal conductivity for some typical materials. Considerable variation can exist amongst nominally identical samples, and careful characterization of defects is needed. The actual temperature dependences of the different regions reflect the interactions of the phonons with other excitations or defects in the crystal. In the absence of these interactions, heat would travel through the crystal with the group velocity of the corresponding lattice waves, and the thermal conductivity of the sample would be limited only by scattering at the surface. We noted earlier that the thermal phonon frequency corresponding to a temperature of 4.2 K is 2.6×10^{11} Hz, and that there is no physical difference between low-energy thermal phonons and high-frequency acoustic waves. The thermal conductivity, κ, is defined operationally by the formula

$$\dot{Q} = -\kappa A \nabla T \qquad (2.22)$$

where $\dot{Q}$ is the energy flow per second perpendicular to an area A, and ∇T is the temperature gradient. κ is in general a second-rank tensor, but until recently the anisotropy of κ had been largely ignored. We wish to obtain an expression for κ in terms of basic physical processes and we begin by noting that there must be a non-equilibrium distribution $n'(\mathbf{q})$ of phonons of wavevector, $\mathbf{q}$, in the x direction, in order that heat should flow in that

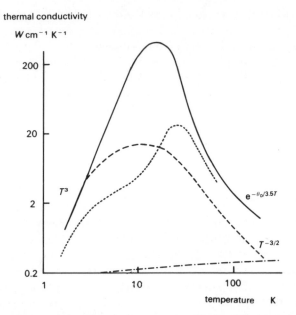

thermal conductivity

Figure 2.6 The measured thermal conductivity of some selected non-metallic solids.
————— purest NaF (Jackson and Walker, 1971) showing Umklapp scattering at higher temperatures
– – – – normal germanium (Geballe and Hull, 1958) showing isotope scattering at higher temperatures
·············· $Al_2O_3 : Ni^{3+}$ (Locatelli and de Goër, 1974) showing resonant scattering by magnetic point defects.
–·–·–·– silica glass (Zeller and Pohl, 1971)

direction. Of course, the equilibrium distribution, $n(\mathbf{q})$, the mean value of which is given by (2.7), cannot give rise to any net heat flow. The heat current, $\dot{Q}$, is given by

$$\dot{Q} = \sum_q n'(\mathbf{q})\hbar\omega(\mathbf{q})v_x(\mathbf{q}) \qquad (2.23)$$

where $\omega(\mathbf{q})$ is the frequency of the phonon having wavevector, $\mathbf{q}$ (for the moment the continuum approximation is not made) and $v_x(\mathbf{q})$ is the x component of the group velocity. We now make the *relaxation time approximation* (embraced also in many areas of physics, not only in phonon and electron scattering), and assume that $n'(\mathbf{q})$ relaxes back to $n(\mathbf{q})$ according to

$$\frac{\partial n'(\mathbf{q})}{\partial t} = \frac{n'(\mathbf{q}) - n(\mathbf{q})}{\tau(\mathbf{q})} \qquad (2.24)$$

This is essentially a Boltzmann transport equation for the phonons, in which $\tau(\mathbf{q})$ is the relaxation time for phonons of wavevector $\mathbf{q}$. For the details of the solution of (2.24) see Klemens (1958) or Ziman (1960). The result is obtained for κ

$$\kappa = \tfrac{1}{3}\sum_{q,p} C_V(\mathbf{q}, p)|\mathbf{v}(\mathbf{q}, p)|\tau(\mathbf{q}, p) \qquad (2.25)$$

where $C_V(\mathbf{q}, p)$ is the heat capacity of phonons of wavevector $\mathbf{q}$ and polarization p. The similarity between (2.25) and the expression given by the simple kinetic theory, $\tfrac{1}{3}C_V v^2\tau$, for the thermal conductivity of a gas is not accidental. There are many phenomena into which the model of a solid as a gas of phonons gives valuable insight; see §2.6. The quantities $C_V(\mathbf{q}, p)$ and $\mathbf{v}(\mathbf{q}, p)$ can be calculated to a fair degree of accuracy, as described above in §2.3. But phonons of each wavevector relax back to $n(\mathbf{q})$ with their own characteristic $\tau(\mathbf{q})$, and this may contain several different contributions. Thus in order to understand the thermal conductivity of a crystal it is necessary to have a knowledge of the magnitude and frequency dependences of all the significant mechanisms by which phonon scattering is taking place. This information is not usually available and, at best, only a qualitative fit can be made to the observed form of $\kappa(T)$. For low-temperature measurements, it is helpful again to make the Debye approximation. It is then possible, using equation (2.18), to calculate the dominant phonon frequency (i.e. that at the peak of the phonon distribution), so that the frequency dependences of scattering processes may be translated directly, if approximately, into the temperature dependence of the thermal conductivity. Where there are several different mechanisms contributing to the total scattering, it can usually be assumed that the scattering rates can be added. Thus

$$\tau^{-1}_{\text{total}}(\mathbf{q}) = \tau_1^{-1}(\mathbf{q}) + \tau_2^{-1}(\mathbf{q}) + \cdots \qquad (2.26)$$

Note, however, that this rule supposes each process to be unmodified by the presence of any of the others, an assumption which, unfortunately, is not always true in practice.

It should be clear by now that thermal conductivity is a much more complicated property than heat capacity. Good accounts of the framework of the subject have been given by Klemens (1958), Callaway (1959), and Berman (1976). The emphasis of research in the last twenty years has been on attempting to understand the detailed mechanisms of phonon scattering, as described below in §2.5. In the earlier preoccupation with $\tau(\mathbf{q})$, however, an interesting property of $|\mathbf{v}(\mathbf{q})|^2$ in equation (2.25) went unnoticed until 1970. It had always been assumed that $\mathbf{v}(\mathbf{q})$ would be fairly isotropic, but Taylor, Maris and Elbaum (1969) observed in heat pulse experiments at helium temperatures that the energy flow of an isotropic (in terms of wavevector) phonon distribution could be strongly concentrated along certain crystal axes. In lithium fluoride, for example, fast transverse energy flow along a [110] direction is 40 times greater than that along a [100] axis. The effect is illustrated in Fig. 2.7. For low-frequency phonons, calculation of the anisotropic energy distribution is straightforward, being determined simply by

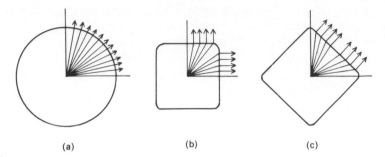

(a)   (b)   (c)

Figure 2.7   Illustration of the manner in which phonon focusing arises from the anisotropy of the surfaces of constant frequency in **q**-space. (*a*), isotropic solid—no focusing; (*b*) and (*c*), phonons focused either parallel, or at 45° to the cubic axes depending on the relative magnitudes and signs of elastic constants. (After Elbaum, 1972.)

Figure 2.8   A map of energy flow in germanium due to phonon focusing. The pattern is observed by a heat-pulse technique in which the position of the heater can be scanned by means of a laser (Northrop and Wolfe, 1980).

the elastic constants of the material. With the use of heat-pulse and superconducting tunnel-junction techniques, however, phonons much further out in the first Brillouin zone can be reached, for which the phase and group velocities are no longer either equal, or always known. Studies of phonon focusing in this regime can give information on the dispersion and scattering of phonons that is not accessible by other methods. Figure 2.8 shows a map of the energy flow in germanium obtained by scanned laser excitation of a thin metal heater and detection with a fixed bolometer (Northrop and Wolfe, 1980). Phonon focusing has its greatest influence on thermal conductivity when the relaxation time of the phonons is long so that they travel ballistically; that is, without significant scattering. But the effect of phonon focusing will always be present to some extent and must be taken into account in any quantitative comparison of theory with experiment for the magnitude of κ.

Phonon focusing is only one example of the application of the heat-pulse technique, which is essentially a pulsed thermal conductivity experiment. Instead of a steady-state heat flow being set up, single, short pulses of heat each lasting about 100 ns are sent through the sample. The phonons making up a heat pulse have a broad thermal distribution in frequency, but since different polarizations travel with different velocities, thoughtful positioning of heater and detector on the sample enables different wavevectors to be selected for study. A review of the principles has been given by von Gutfeld (1968). Figure 2.9(a) shows the detected heat pulses that have travelled through sodium fluoride. The sharpness of the individual polarization modes indicates that little scattering is taking place within the body of the sample. In contrast, Fig. 2.9(b) shows diffusive heat flow in a sample of magnesium oxide, where the ballistic modes are almost lost in the scattered radiation reaching the detector. Because of its ability to resolve the phonon modes, the heat-pulse technique is a valuable adjunct to thermal conductivity in studying phonon scattering.

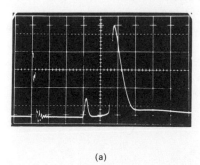

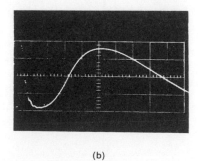

(a)   (b)

Figure 2.9 (a) Ballistic heat pulses in sodium fluoride. The longitudinal and transverse modes are clearly separated because of the very low scattering (Rogers, 1971). (b) Heat pulses in ferrous doped magnesium oxide. The ballistic propagation is almost totally removed by resonant and Rayleigh scattering of phonons by the ferrous ions (Wigmore, 1971a).

2.5 Scattering of phonons

We begin by considering the scattering of phonons in the perfect model solid at the lowest temperatures. Since phonons are normal modes of the ideal harmonic lattice, they are free (in the absence of any other excitations) to travel through such a crystal totally without loss until they collide with a surface and undergo boundary scattering. Casimir (1938) suggested that all the phonons striking a surface would be diffusely reflected by random surface irregularities, which he guessed would be large compared with the phonon wavelengths. The mean free path of the phonons would then simply be the dimension of the sample, so that $\tau(\mathbf{q})$ should be independent of phonon frequency. The only temperature-dependence of κ would enter (2.25) through $C_V(\mathbf{q}, p)$, which, as we showed earlier, is proportional to T^3.

The T^3 temperature variation of κ is indeed widely observed amongst non-metals at low temperatures. However, other experiments show that the diffusive scattering proposed by Casimir cannot be a complete picture of what happens to the phonons. There is evidence from both thermal conductivity (Thacher, 1967) and heat-pulse experiments (Wigmore, 1971b) that the roughened surface of a crystal may conceal a subcutaneous layer of badly dislocated material extending as far as 0.2 mm below the visual surface. Instead of being diffusively scattered, the incident phonons are actually absorbed in the layer and others re-emitted in a different thermal distribution corresponding to the new local temperature. The focus of attention, then, has shifted to the interaction of phonons with dislocations and other defects in crystals. Quite apart from the intrinsic physical interest, these phenomena are relevant to the current technical problem of producing perfect, defect-free substrates for epitaxial microelectronic devices.

As the temperature of the solid is raised, so does the amplitude of vibration of each atom increase in its potential well. The motion can no longer be modelled as a simple harmonic oscillator, since the higher-order terms in the lattice potential (2.1) give additional contributions to the restoring force which are not proportional to the displacement of the atom. However, these effects are still relatively small and it is possible to treat the anharmonic terms as a perturbation which causes interactions between the harmonic phonons. It is worth bearing in mind, however, that strictly the lattice vibrations of §2.2 are then no longer pure eigenstates of the crystal lattice.

In order to understand qualitatively the effect of the anharmonicity we rewrite (2.1) a little differently as

$$F = \frac{C'}{2!}S^2 + \frac{D'}{3!}S^3 \qquad (2.27)$$

where F is now the free energy per unit volume of the crystal, S is the strain and

C' and D' are constants. In general, the stress, T, due to a given strain is

$$T = \frac{dF}{dS} = C'S + \frac{D'}{2}S^2 \qquad (2.28)$$

(The quantity C' is the elastic constant of the material.) If the total strain at a point is due to the simultaneous presence of two phonons

$$S = S_1 \exp i\{\omega_1 t + q_1 x\} + S_2 \exp i\{\omega_2 t + q_2 x\} \qquad (2.29)$$

then the resulting stress will contain not just $\exp i\{\omega_1 t + q_1 x\}$ and $\exp i\{\omega_2 t + q_2 x\}$, but also a component $\exp i\{(\omega_1 + \omega_2)t + (q_1 + q_2)x\}$. Thus the anharmonic term $(D'/2)S^2$ has coupled the two original lattice waves together to generate a third, which has frequency and wavevector given by

$$\omega_1 + \omega_2 = \omega_3 \qquad (2.30)$$

and

$$q_1 + q_2 = q_3 \qquad (2.31)$$

These two expressions are often presented as conservation rules governing the collision of two phonons, (2.30) expressing the conservation of energy, and (2.31) the conservation of momentum. However, $\hbar q$ actually represents crystal momentum, not true inertial momentum. In order to emphasize that a phonon wavevector is physically meaningful only if defined in the first Brillouin zone, the conservation of crystal momentum is usually written

$$q_1 + q_2 = q_3 = g + q_3' \qquad (2.32)$$

where g is a reciprocal lattice vector, that is, in one dimension, a multiple of $2\pi/a$, and q_3' is a phonon in the first Brillouin zone.

Physically, a most important distinction exists between phonon collisions in which $g = 0$, termed *normal processes*, and those for which the $q_1 + q_2$ is beyond the edge of the first Brillouin zone. The latter are known by the unforgettable name of *Umklapp processes* from the German for 'to flop over'. We first discuss normal processes. As far as heat flow in the crystal is concerned, normal processes make no direct contribution to the relaxation of the non-equilibrium distribution in (2.23). They may change the distribution of the energy amongst the different modes but, because energy and momentum are directly conserved, there is no overall resistance to flow. Therefore, the relaxation rate relating to the normal processes cannot simply be added to other resistive terms, using (2.26). However, normal processes cannot be ignored since they may have the effect of redirecting energy into a new set of lattice modes that can relax faster than the original distribution. From this point, the theory depends on the details of scattering in particular materials, and especially on the relative strengths of normal processes and other scattering. Further details may be found in the article by Callaway (1959) or in the book by Berman (1976). One of the most useful techniques for studying normal phonon scattering, albeit at rather lower phonon frequencies, is ultrasonic attenuation, and a typical train of ultrasonic echoes used in such investigations, at a frequency of 10 GHz, is shown in Fig. 2.10. The non-exponential envelope is due to the very slight deviation of the sample from perfect alignment and parallelism. Thermal phonon frequencies are rather higher than 10 GHz, but ultrasonic data can be characterized so precisely as to frequency, polarization and wavevector that the results obtained by this technique can be convincingly scaled. The ultrasonic wave is usually generated from a microwave electric field through the piezoelectric effect. It is an important feature that the acoustic phonons are generated coherently. Normal process scattering of the phonons will destroy the phase coherence and hence be observed as attenuation even if total energy flow and crystal momentum are conserved. Maris (1971) has reviewed the study of normal phonon–phonon scattering through its effect on ultrasonic attenuation. Although details rapidly become very complex, the physical principle is straightforward and instructive. Consider the low dispersion curves for the acoustic phonons along a pure mode axis of a typical solid, shown in Fig. 2.11. If the relation for each polarization mode were exactly a straight line, then ultrasonic phonons of any polarization could scatter by a collinear process of the type, for example,

$$L_u + L \rightarrow L$$

meaning that an ultrasonic longitudinal phonon scatters from a thermal

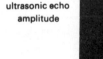

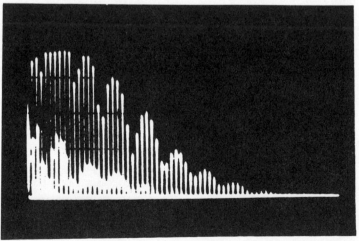

Figure 2.10 Ultrasonic echoes observed in quartz at a frequency of 10 GHz. Each pulse corresponds to a further round trip in the sample of 20 mm. The modulation of the amplitudes is due to slight misalignment and non-parallelism of the sample.

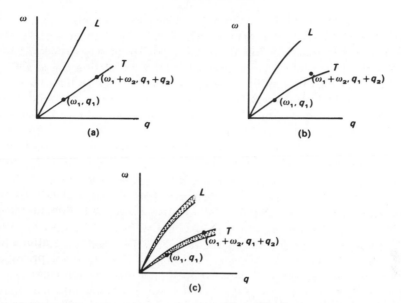

Figure 2.11 Illustration of the manner in which the normal phonon scattering process $T_u + T \rightarrow T$, forbidden in a dispersive crystal, is made possible by the finite lifetimes of the phonons. (a) no dispersion: phonons (ω_1, q_1) and $(\omega_1 + \omega_2, q_1 + q_2)$ both lie on the dispersion curve; (b) with dispersion: $(\omega_1 + \omega_2, q_1 + q_2)$ is not an allowed phonon state; (c) with the broadening due to the finite lifetime of phonons: $(\omega_1 + \omega_2, q_1 + q_2)$ becomes allowed again.

longitudinal phonon to produce a second thermal longitudinal phonon of higher frequency. It is a simple geometrical exercise to show that besides the above process the only others that satisfy the conservation rules (2.30) and (2.31) are

$$T_u + T \rightarrow T$$

and

$$T_u + L \rightarrow L$$

The last of these is not a collinear process. In a real solid, however, dispersion means that the $\omega - q$ plots are actually curved. It is then impossible to satisfy the conservation rules either for $L_u + L \rightarrow L$ or for $T_u + T \rightarrow T$, as a little more geometry will quickly show. This was an obstacle in the theory of normal phonon scattering for many years, since experimentally the attenuation of longitudinal ultrasound was found to be just as large as that of transverse waves. Finally, Maris (1964) and Simons (1963) showed that the Heisenberg uncertainty principle was responsible for relaxing the conservation rules. Since the various phonons in reality have finite lifetimes, their frequencies are uncertain to the extent of the reciprocals of their lifetimes. These uncertainties are sufficient to reinstate the processes disallowed by dispersion. In general, the phonon attenuation due to normal processes varies roughly as ωT^n, where n has a value somewhere between 4 and 9 depending on the material and the phonon polarization. It is interesting to note that, in the particular case of liquid ^{4}He the curvature of the phonon dispersion relation can actually be varied by externally applied pressure, so that the magnitude of normal processes can be adjusted at will. In §2.6 we shall describe two other interesting phenomena for which normal processes are important, that is, the Poiseuille flow of phonons, and second sound in solids.

In contrast to normal scattering, Umklapp processes present a direct resistance to the flow of heat. If the sum of q_1 and q_2 is greater than π/a, (but not of course, greater than $2\pi/a$), then it is clear from the dispersion curve in Fig. 2.2 that its group velocity must be negative. Another way of expressing this idea is through (2.32), which reduces q_3 to its equivalent wavevector in the first Brillouin zone. The process is exactly like that of Bragg reflection of an X-ray or electron from a crystal lattice. However it is described, the result is that, surprising as it may seem, the energy flow of q_3 is in the opposite direction to that of q_1 and q_2. At very low temperatures, there are no phonons of sufficiently large wavevector in the crystal to give a q_3 beyond the boundary of the first Brillouin zone, so that Umklapp scattering is negligible. It may be shown on the basis of the Debye model that the probability of the two scattered phonons having wavevectors greater than about a half of the value at the zone edge is approximately proportional to $\exp(-\theta_D/T)$. Experimentally, an exponential dependence is indeed observed in a few materials, although the exponent often disagrees with the value predicted by this simple argument by a factor of up to 3 (Jackson and Walker, 1971). In many solids, however, the effect of Umklapp processes is not observed at all because of the dominance at higher temperatures of other forms of phonon scattering due to defects.

Many of the most important properties of solids, in terms of their applications, have their origins in defects of various kinds. Measurements of phonon scattering by the defects can provide valuable information with which to characterize the scattering centres themselves. Indeed, there are some impurities which are so weakly coupled to electromagnetic radiation, as for example in optical spectroscopy or electron spin resonance, that they can be studied only through the use of phonon techniques. Such measurements are usually best made at low temperatures in order to minimize the effects of phonon–phonon scattering. We have already discussed thermal conductivity at some length, and mentioned heat pulses and ultrasonic attenuation. Ultrasound can be used only up to frequencies of about 10 GHz, but the ultrasonic phonons have the particular advantage of being monochromatic, coherent and well defined in polarization and wavevector. Thermal techniques, on the other hand, specify the phonon only to within a broad distribution, but have the advantage that the frequency can be made much higher, approaching the edge of the Brillouin zone. Another particularly valuable technique is based on the use of superconducting tunnel junctions as phonon generators. Eisenmenger (1976) has reviewed single-particle tunnel-ling devices, which allow phonons in the frequency range from about 100 GHz to greater than 1000 GHz to be studied, with resolution both of frequency and polarization. More recently Berberich *et al.* (1982) have demonstrated that resolution as fine as MHz can be achieved using Josephson junctions as generators. The physical principles of these devices will be described in chapter 8; we show in Fig. 2.12 the spectrum of phonons emitted by a single-particle junction at temperatures well below T_c.

The scattering of phonons by point defects in crystals, such as impurities or vacancies (as opposed to larger entities, such as dislocations) can conveniently be subdivided into scattering by non-resonant point defects, and by those that possess internal transitions which may be excited by the phonons. The primary effect of introducing a foreigner into the lattice is to break the translational symmetry on which the vibrational mode analysis of §2.2 was based. Either the mass, m, or the force constant, F, or both, will be modified. The effect on a lattice wave of the differing potential is analogous to the Rayleigh scattering of light by small particles in a transparent medium—the origin of the blue sky. The scattering is elastic; that is, the frequency of the wave is not changed, only its direction. A simple argument shows that for Rayleigh scattering

$$\tau^{-1}(\mathbf{q}) \propto |\mathbf{q}|^4 \tag{2.33}$$

and Klemens (1958) verified that the same expression holds also for phonons. Substituting this into (2.25) yields eventually a thermal conductivity that varies with temperature as $T^{-3/2}$. Very many materials show this effect because, quite apart from impurities introduced during crystal growth, they

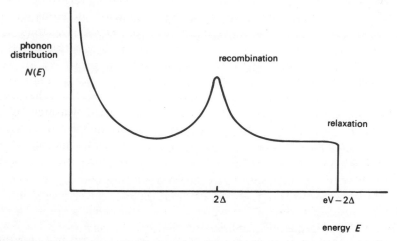

Figure 2.12 The spectrum of phonons emitted by a superconducting tunnel junction at temperatures well below T_c. Only the relaxation phonons are shown since these are the ones used in the tunable phonon spectrometer (Eisenmenger, 1976).

contain a mixture of isotopes. It is possible for atoms in different unit cells of a crystal, although indistinguishable to X-rays, actually to differ by a neutron or two in the nucleus. This may be a small percentage in terms of atomic mass but the concentration of the isotope may be as high as 50%. Isotope scattering can be large enough to dominate completely the thermal conductivity of a material at all but the lowest temperatures and often, in order to study other sources of scattering such as Umklapp processes, it is necessary to work in isotopically pure crystals.

The resonant interaction of phonons with defects is an area of great current interest, and the reader is referred to recent reviews of phonon spectroscopy by Bron (1980), and by Challis (1983). Although all defects are affected to some extent by lattice vibrations, those centres that display the Jahn–Teller effect (see §3.1) figure most prominently because of their very strong interaction between electrons and phonons. Tiny amounts of, for example, chromium or iron impurities, too small to be detected by techniques using photons, can modify drastically the magnitude and temperature dependence of the thermal conductivity. The different models of the defects can be tested through the sets of energy levels that they predict.

2.6 Solid helium—a quantum crystal

Crystals of solid helium are very few and far between mainly because, as discussed in §1.5, they cannot be made simply by cooling liquid helium to lower temperatures. It is necessary to apply pressures of at least 25 atmospheres in order to solidify even ^{4}He, and the high-temperature phase diagrams of the two isotopes show that several different phases may be found under various conditions.

The interest in these two materials, from a phonon standpoint, is that the amplitudes of the vibrations of the atoms in their potential wells are extremely large, so large in fact that many of the basic assumptions of the harmonic model of lattice vibrations described earlier are simply not valid. As we saw in §1.5, the magnitude of the vibrations is a direct consequence of the quantum nature of matter. It will be recalled (2.6) that, according to the quantum theory, the eigenstates of a harmonic oscillator are $(n + \frac{1}{2})\hbar\omega$. For 'normal' solids, the value of n, which is given by Bose–Einstein distribution (2.7), is a lot larger than 1/2. The thermal excitation of the lattice vibrations dominates the quantum mechanical zero-point motion. In solid helium, the situation is reversed so that, even at absolute zero, the root mean square displacement of an atom is more than 25% of the distance to its nearest neighbour. As previously discussed, it is this large zero-point kinetic energy that prevents the solid phase being formed at all unless considerable external pressure is applied. The other noble gas solids, neon, argon, krypton and xenon, also exhibit the effects of zero-point motion, but to a much lesser extent. Guyer (1969) has reviewed the physics of quantum crystals in considerable detail.

Because of the large amplitude of atomic vibration, the anharmonicity of the lattice plays a much more important role than in normal solids. Indeed, if the attempt is made to set up the Taylor series (2.1) for the potential energy of an atom, it is found that the cubic coefficient D is at least as large as the harmonic C, so that the series does not converge. Not only is the harmonic approximation not valid, but even the Taylor series itself cannot be used. Can then the excitations of solid helium be regarded as phonons at all? Reassurance is found from heat-capacity measurements which show the T^3 law characteristic of harmonic phonons with θ_D of 26 K (^{4}He). However, the harmonic oscillators to which the phonons refer are not now the usual normal modes of lattice vibration and no long-range coupled motion of the atoms occurs. The model of solid helium is closer to that of the Einstein solid with each atom vibrating individually. The potential well of the oscillator must be taken, however, not as a static function as with a normal crystal, but as a time average of the motion of one atom relative to its neighbours. Since the potential determines the motion of the atoms, which in turn defines the potential, the phonons excited are called *self-consistent phonons*.

The large anharmonicity of solid helium also gives rise to interesting effects in phonon scattering. Phonon–phonon interactions are necessarily strong and by carrying out experiments at temperatures below 1 K or so, it is possible to freeze out the Umklapp processes leaving the normal processes to be studied in relative isolation. Under these conditions a new mode of heat transport is observed; this is the *Poiseuille flow* of phonons. We have referred more than once to modelling a solid as a gas of phonons, and we can regard the flow of phonons along a crystal as analogous to the flow of gas along a pipe. Knudsen flow, in which the only scattering of the gas molecules takes place at the walls of the pipe, is clearly the parallel of thermal conductivity in the boundary scattering regime and is observed in many crystals. In the Poiseuille limit, on the other hand, the dominant resistance to flow is due to scattering of molecule by molecule, or of phonon by phonon, with energy and momentum being conserved and in consequence with no net resistance to flow. In most solids, the anharmonicity is so weak that, as the normal processes increase in number with increasing temperature, so also do the resistive Umklapp processes. Only in solid helium and, arguably, in some alkali halides, has Poiseuille flow actually been observed. Figure 2.13 demonstrates that the mean free path of the phonons in solid ^{4}He in the Poiseuille regime is larger than the size of the sample, which would be its limiting value under Knudsen conditions.

A closely related phenomenon is that of *second sound*. Just as an ordinary sound wave travels in a gas comprised of molecules, so can a density wave propagate through a gas of phonons. Since phonon density depends on temperature, second sound corresponds macroscopically to a temperature wave. Ward and Wilks (1951) first showed that such a disturbance should travel at a velocity $1/\sqrt{3}$ of the normal, first-sound velocity, and this result has been confirmed by many experiments. The experimental conditions for its

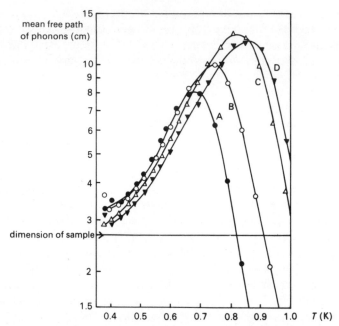

Figure 2.13   The enhanced phonon mean free path in solid ⁴He due to Poiseuille flow. The dimension of the sample would be the mean free path of the phonons in the Knudsen limit. The different curves correspond to different sample orientations (Hogan *et al.*, 1969).

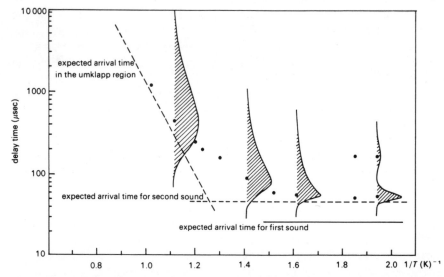

Figure 2.14   Illustration of second sound in solid ⁴He. As the temperature is lowered (towards the right of the diagram) the velocity of the observed heat pulse approaches asymptotically that expected for second sound. At higher temperatures, the heat flow becomes diffusive due to the increasing contribution to phonon scattering from Umklapp processes (Ackerman *et al.*, 1966).

existence are similar to, but not quite as restrictive as those for Poiseuille flow. In addition to solid ^{3}He and ^{4}He, second sound has been seen in lithium fluoride, sodium fluoride, and bismuth. Figure 2.14 shows the development of a second sound heat pulse in ^{4}He as the temperature is varied (Ackerman *et al.*, 1966).

2.7 Related topics of current interest

In this section, we describe briefly three additional low-temperature topics that are connected with lattice vibrations.

First we discuss the phenomenon of thermal expansion. It is a fact of considerable importance in the design of low-temperature equipment that constructional materials contract by anything up to 1% between room and liquid helium temperatures. The effect is a direct consequence of the anharmonic terms in the lattice potential (2.1). At the lowest temperatures the atoms stay close to the symmetric minima of their potential wells. As their thermal energy increases, because of the anharmonic asymmetry of the wells, they spend on average more time at greater separations. It is not difficult to estimate the increase, $\bar{x}$, in separation. Assuming a Boltzmann distribution for the lattice vibrations and restricting consideration to one dimension, we can write

$$x = \frac{\int\limits_{-\infty}^{+\infty} x e^{-V(x)/k_B T} dx}{\int\limits_{-\infty}^{+\infty} e^{-V(x)/k_B T} dx} \tag{2.34}$$

where $V(x)$ is (2.1) with x written for $\mathbf{r} - \mathbf{r}_0$.

On carrying out the integrations (Kittel, 1976) one obtains

$$\bar{x} = \frac{3D}{4C^2} k_B T \tag{2.35}$$

where C and D are coefficients in the expansion of (2.1). The linear coefficient of thermal expansion, α, is more often written in terms of a quantity γ, called the Grüneisen constant, which characterizes the way the lattice vibrational frequencies ω change with volume V, through

$$\frac{\Delta\omega}{\omega} = \gamma \frac{\Delta V}{V} \tag{2.36}$$

Then α is given by

$$\alpha = \frac{1}{3}\left(\frac{\partial V}{\partial T}\right)_P = \frac{\gamma C_V}{3V\eta} \tag{2.37}$$

Since η, the bulk modulus, does not change very much with temperature, α is proportional to C_V, the heat capacity, varying with the temperature, for example, as T^3 at low temperatures. Because thermal expansion depends on spatially averaged properties of the lattice, it is a less sensitive probe of the anharmonicity than some other techniques, such as ultrasonic attenuation. Nonetheless, it is clearly a phenomenon of great technological impact.

Secondly, we discuss in an introductory manner the thermal properties of glasses. Such materials are much less well understood than crystalline solids, and a great deal of work has been done in recent years—and is still being done—on insulating, metallic, and semiconducting glasses. The arrangement of atoms in an amorphous solid has a large measure of short-range order, that is to say, each atom is surrounded by the right number of neighbours in roughly the right directions, but there is no long-range correlation of positions. How then is it possible to conceive of lattice vibrations where there is no lattice? The interested reader is directed to the reviews by Böttger (1974), and by Weaire (1981) for further enlightenment, since a full discussion would be out of place here. Briefly, the answer to the question is that thermal and acoustic properties depend on the existence of a distribution of harmonic vibrational states, to some extent regardless of the actual details of the vibrations. It can be shown that the low-frequency vibrations in particular are determined by the short-range order of the glass, which is approximately crystal-like. Thus it is to be expected that at low temperatures both the heat capacity and the thermal conductivity of glasses will show the crystalline T^3 dependence.

Figure 2.15 shows the measured variation of these properties with temperature in silica, a typical glass. It is found that at temperatures below about 1 K, an additional contribution to the heat capacity starts to appear, and that by a temperature of 20 mK, the heat capacity of a glass is a thousand times larger than that of the corresponding crystal. The additional term in the heat capacity is found to be linearly proportional to temperature. In the same temperature range, the thermal conductivity of silica tends towards a T^2 dependence. To the extent of the dominant phonon approximation (§2.4), such a dependence indicates a phonon mean free path, λ, proportional to ω^{-1}, since $\kappa \propto C_V \lambda$, $C_V \propto T^3$, and $\omega \propto T$. The surprising thing is that similar effects both in magnitude and in temperature dependence are observed in almost all glasses, regardless of composition or method of production. The inescapable conclusion is that, in addition to the phonons, there are some other excitations, as yet unidentified, present in glasses which contribute to their specific heat and phonon scattering. Because they occur so widely, these excitations must be associated with the amorphous structure itself, and it has been clearly established that they are not due to defects. In the early 1970s, it was realized that the observed dependences, $C_V \propto T$ and $\lambda \propto \omega^{-1}$, are characteristic of the behaviour of a resonant spin or other two-level system consisting just of two states interacting with resonant radiation. The actual resonant frequencies must be spread uniformly over a wide distribution in order to explain the

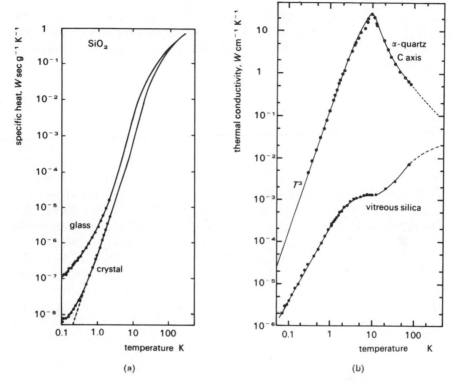

Figure 2.15 (a) The heat capacity, and (b) the thermal conductivity of vitreous (amorphous) silica, compared with the data for crystalline quartz (Zeller and Pohl, 1971).

scattering of thermal phonons. The microscopic origin of these two-level systems is not yet understood. In the present state of knowledge, the most widely held theory is that they are energy splittings caused by atoms in the amorphous structure tunnelling between similar configurations; but ideas are changing rapidly in this area. Phillips (1981) has edited an excellent and readable review of the field.

Finally, we introduce an unsolved problem that is of central importance in experimental low-temperature physics, *the Kapitza boundary resistance*. Although seemingly trivial, this anomaly has survived the attentions of a large number of experimenters and theoreticians over the last forty years. The problem is this. At a boundary between two dissimilar materials there must always be a temperature discontinuity, ΔT. This follows from the fact that heat flow is carried by phonons, of which only a fraction will be transmitted, while the remainder are reflected. The phenomenon is characterized in terms of a thermal boundary resistance, the Kapitza resistance, defined by

$$R_K = \Delta T / \dot{Q} \qquad (2.38)$$

Since phonons are quanta of lattice vibrations, it should be possible to calculate R_K from a knowledge of the acoustic impedances of the two materials together with an average over the anisotropic phonon distribution in the two materials. In the very simplest case of normal incidence the problem is simply that of matching transmission lines with

$$T = \left(\frac{2Z_1}{Z_1 + Z_2} \right)^2 \qquad (2.38)$$

where T is the transmission coefficient, and Z_1 and Z_2 are the acoustic impedances of the two media. Herth and Weis (1970) describe the full three-dimensional details of the calculation, which gives reasonable agreement with the measured transmission, *except* when one of the two media is liquid helium.

Whether the isotope is ^{3}He or ^{4}He, the measured thermal transmission coefficient exceeds that calculated above by a factor of anything between 5 and 500. Experiments have been carried out on rough, smooth, polished, etched, and cleaved surfaces. The anomaly may differ in magnitude but it always remains, although the disagreement with theory becomes less as the temperature is reduced. In addition to the measurement of the thermal conductivity, most of the techniques described in §2.5 have been focused on the problem. Heat pulses have been used to measure transmission coefficients of individual modes (Guo and Maris, 1974). The frequency dependence of the effect has been measured using superconducting tunnel junctions (Kinder and Dietsche, 1974). The angular distribution of phonons emitted into liquid helium from a cleaved crystal of sodium fluoride has been determined in a series of experiments reviewed by Wyatt (1980).

Over the years, various mechanisms have been proposed to explain the discrepancy, but so far without quantitative success. An early model proposed that better acoustic matching could be accomplished, as with the 'blooming' of lenses, through a dense layer of solid helium being formed at the solid–helium interface (Challis et al., 1961). Enhancement of phonon transmission due to mechanical resonances of surface impurities (van der Sluijs et al., 1974), and the phonon-induced desorption of helium atoms from the layer of solid helium at the boundary (Long et al., 1974), are similarly unable to explain the details of the Kapitza resistance. The most recent theory, by Shiren (1981) seeks to explain the breakdown of the acoustic mismatch model in terms of surface roughness in conjunction with the generation of surface acoustic waves. The most useful recent reviews of the topic are those by Challis (1974), Kinder et al. (1980), and Wyatt (1980).

In conclusion, it should be emphasized that the Kapitza resistance is much more than merely an esoteric oddity. The efficiency of heat transfer is of crucial importance in reaching and maintaining low temperatures, and particularly in the attempt to reach the millikelvin regime good thermal contact can make the difference between success and failure.

Bibliography

References

Ackerman, C.C., Bertman, B., Fairbank, H.A. and Guyer, R.A. *Phys. Rev. Lett.* **16**, 789 (1966).
American Institute of Physics Handbook, American Institute of Physics, New York (1963).
Berberich, P., Buemann, R. and Kinder, H. *Phys. Rev. Lett.* **49**, 1500 (1982).
Casimir, H.B.G. *Physica* **5**, 495 (1938).
Challis, L.J., Dransfeld, K. and Wilks, J. *Proc. Roy. Soc.* **A260**, 31 (1961).
Clement, J.R., and Quinnell, E.H. *Phys. Rev.* **92**, 258 (1953).
Egelstaff, P.A. (ed.) *Thermal Neutron Scattering*. Academic Press, New York (1965).
Einstein, A. *Ann. Physik* **22**, 180 (1907).
Elbaum, C., in *Proc. Int. Conf. on Phonon Scattering in Solids*. Ed. H.J. Albany, Paris 1 (1972).
Geballe, T.H. and Hull, G.W. *Phys. Rev.* **110**, 773 (1958).
Guo, C.J. and Maris, H.J. *Phys. Rev.* **A10**, 960 (1974).
Herth, P. and Weis, O. *Z. Angew. Phys.* **29**, 101 (1970).
Hogan, E.M., Guyer, R.A. and Fairbank M.A. *Phys. Rev.* **185**, 356 (1969).
Jackson, H.E. and Walker, C.T. *Phys. Rev.* **B3**, 1428 (1971).
Kinder, H. and Dietsche, W. *Phys. Rev. Lett.* **33**, 578 (1974).
Locatelli, M. and de Goër, A.M. *Solid St. Comm.* **14**, 111 (1974).
Long, A.R., Sherlock, R.A. and Wyatt, A.F.G. *J. Low Temp. Phys.* **15**, 523 (1974).
Maris, H.J. *Phil. Mag.* **9**, 901 (1964).
Maris, H.J. *Physical Acoustics*. Ed. W.P. Mason and R.N. Thurston, Academic Press, New York, **8**, 279 (1971).
Northrop, G.A. and Wolfe, J.P. *Phys. Rev.* **B22**, 6196 (1980).
Rogers, S.J. *Phys. Rev.* **B3**, 1440 (1971).
Shiren, N.S. *Phys. Rev. Lett.* **47**, 1466 (1981).
Simons, S. *Proc. Phys. Soc.* **82**, 401 (1963).
Taylor, B., Maris, H.J. and Elbaum, C. *Phys. Rev. Lett.* **23**, 416 (1969).
Thacher, P.D. *Phys. Rev.* **156**, 975 (1967).
Van der Sluijs, J.C.A., Jones, E.A. and Alnaim, A.E., *Cryogenics* **14**, 95 (1974).
Ward, J.C. and Wilks, J. *Phil. Mag.* **42**, 314 (1951).
Wigmore, J.K. *J. Physique* **C1**, 766 (1971a).
Wigmore, J.K. *Physics Lett.* **37A**, 293 (1971b).
Woods, A.D.B., Brockhouse, B.N. Cowley, R.A. and Cochran, W. *Phys. Rev.* **131**, 1025 (1963).
Zeller, R.C. and Pohl, R.O. *Phys. Rev.* **B4**, 2029 (1971).

Further reading

Ashcroft, N.W. and Mermin, N.D. *Solid State Physics*. Holt, Rinehart and Winston, New York (1976).
Berman, R. *Thermal Conduction in Solids*. Clarendon Press, Oxford (1976).
Böttger, H. 'Vibrational properties of non-crystalline solids', in *Phys. Stat. Solidi* **62B**, 9 (1974).
Bron, W.E. 'Spectroscopy of high frequency phonons', in *Rep. Progr. Phys.* **43**, 301 (1980).
Callaway, J. 'Model for lattice thermal conductivity at low temperatures', in *Phys. Rev.* **113**, 1046 (1959).
Challis, L.J. 'Kapitza resistance and acoustic transmission across boundaries at high frequencies', in *J. Phys. C (Solid State)* **7**, 481 (1974).
Challis, L.J. 'Phonon spectroscopy', in *Contemp. Phys.* **24**, 229 (1983).
Eisenmenger, W. 'Superconducting tunnel junctions as phonon generators and detectors', in *Physical Acoustics*, eds. W.P. Mason and R.N. Thurston, **12**, 79, Academic Press, New York, (1976).
Guyer, R.A. 'The physics of quantum crystals', in *Solid State Physics*, eds. F. Seitz, D. Turnbull and H. Ehrenreich, **23**, 413, Academic Press, New York (1969).
Kinder, H., Weber, J. and Dietsche, W. 'Kapitza resistance studies using phonon pulse reflection', in *Phonon Scattering in Condensed Matter*, ed. H.J. Maris, Plenum, New York (1980), p. 173.

Kittel, C. *Introduction to Solid State Physics*. 5th edn., Wiley, New York (1976).

Klemens, P.G. 'Thermal conductivity and lattice vibrational modes', in *Solid State Physics, op. cit.*, **7**, 1 (1958).

Leighton, R.B. *Principles of Modern Physics*. McGraw-Hill, New York (1959).

Mandl, F. *Statistical Physics*, Wiley, London (1981).

Mendelssohn, K. *The Quest for Absolute Zero: the Meaning of Low Temperature Physics*. 2nd edn., Taylor and Francis, London (1977).

Musgrave, M.J.P. *Crystal Acoustics*. Holden Day, San Francisco (1970).

Phillips, W.A. (ed.) *Amorphous Solids—Low Temperature Properties*. Springer-Verlag, Berlin (1981).

von Gutfeld, R.J. 'Heat Pulse Transmission', in *Physical Acoustics, op. cit.*, **5**, 233 (1968).

Weaire, D.L. 'The vibrational states of amorphous semiconductors', in *Amorphous Solids, op. cit.*, 13.

Wyatt, A.F.G. 'Characteristics of Kapitza conductance', in *Phonon Scattering, op. cit.*, p. 181.

Ziman, J.H. *Electrons and Phonons*. Clarendon Press, Oxford (1960).

3 Electrons

3.1 Electrons—basic ideas

In this chapter, we shall describe some of the low-temperature properties of solids that occur because of the presence of electrons. The reader is referred to chapter 4 for a discussion of the related topic of superconductivity. Before focusing specifically on low-temperature aspects, however, we shall review some of the relevant concepts of electronic behaviour in solids. As in the description of phenomena arising from lattice vibrations, the states lying immediately above the $T = 0$ ground state play the most important role. By describing solid state phenomena in terms of separate electron and phonon systems, we are implicitly making a very basic assumption about solid state physics called the *Born–Oppenheimer approximation* (Ziman, 1964). Since the masses of the atomic nuclei in a crystal are so much greater than those of the electrons, the Schrödinger equation for the whole crystal is separable into two simpler equations which describe independently the motion of the nuclei and that of the electrons. The total eigenstate of the crystal is a product of the separate nuclear and electronic wavefunctions, and the eigenvalues can be added together to obtain the total energy of the solid. Expressed in this way, the Born–Oppenheimer approximation seems so obviously reasonable as hardly to merit explicit attention. The fact is that the approximation is valid only when the orbital electronic states are non-degenerate, and a whole range of interesting phenomena, known collectively as the *Jahn–Teller effect*, manifest themselves when this condition is not satisfied. We referred in chapter 2 to the consequences of the Jahn–Teller effect for the scattering of phonons. In other materials, for example, the rare earth vanadates, the strong electron–phonon coupling induces low-temperature structural phase transitions, as described by Melcher (1976). For a survey of the consequences of the Jahn–Teller effect, see Engelman (1972).

Whereas the electronic energy levels of individual, isolated atoms are discrete and widely separated in energy, those of solids lie in quasi-continuous bands. Figure 3.1 illustrates schematically the transition from one regime to the other that can be imagined to take place as a large number of isolated

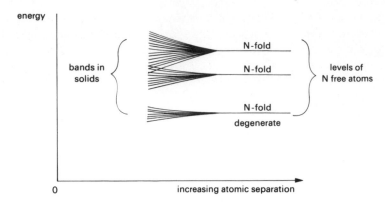

Figure 3.1 Illustration of one approach to the origin of energy bands in solids. As atoms are brought closer together, their mutual interaction causes a divergence of the levels, although the total number of states remains fixed.

atoms are brought together to form a solid. The total number of states within a band is determined solely by the number of atoms. Thus it is the same regardless of whether the width of the band is very narrow, corresponding to large distances between the atoms, or whether it is broadened due to an increase of the interatomic forces as the atoms are brought closer together. The broadening is less for the inner atomic orbitals because the overlap of wavefunctions between different atoms is smaller. On the other hand, the highest-energy bands, corresponding to the outer electrons of the atoms, may be broadened so much that their energies overlap.

The extent to which electrons actually occupy the available states is crucial in determining the physical properties of a particular solid. Since electrons are fermions obeying the Pauli exclusion principle, the states are filled in order of ascending energy, from states in the very lowest bands, up to some maximum, known as the *Fermi level*, with energy E_F. The magnitude of E_F is typically a few electron volts so that, even at room temperature, thermal excitation of electrons does not affect these general conclusions ($k_B T$ is equal to 1 eV for a temperature of 1.16×10^4 K). In semiconductors and insulators, the Fermi level lies in a forbidden energy region between two bands of allowed states. For a metal, E_F lies within an allowed band of energies so that there are many empty states immediately above the Fermi level into which electrons can be excited by, for example, the influence of an electric field, giving rise to electrical conductivity. Indeed, the physical properties of metals at low temperatures are totally dominated by the electrons occupying states close to the Fermi level. It is important, therefore, to learn as much as possible regarding the nature of these states.

Many of the basic experimental phenomena in metals suggest that the allowed electronic states have similarities with the plane wave eigenstates of non-interacting free particles. Note, however, that the mere existence of electrical conduction does *not* prove the presence of free electrons; see §3.5. The free-electron theory has been very successful in providing a first order model for the properties of metals, although, as we shall see, there are many details which cannot be fitted. We give only a brief review; Kittel (1976) and Ashcroft and Mermin (1976) provide fuller coverage.

The solutions of Schrödinger's equation, for free particles in a cubical box of side L, are similar to the lattice vibrational eigenstates in the form of standing waves,

$$\psi_k(\mathbf{r}) = A \sin k_x x \sin k_y y \sin k_z z \tag{3.1}$$

where

$$k_x, k_y, k_z = \pi/L, 2\pi/L, \ldots \infty$$

Convention defines the wavevector of an electron to be $\mathbf{k}$, as opposed to $\mathbf{q}$ for a phonon. Each eigenstate can accommodate only two electrons, with opposite spins. The dispersion relation for the free electrons is then

$$E_k = \frac{\hbar^2}{2m_e}(k_x^2 + k_y^2 + k_z^2) = \frac{\hbar^2}{2m_e}|\mathbf{k}|^2 \tag{3.2}$$

If the solid contains N electrons, they will fill up the eigenstates two by two until the Fermi level is reached. It can be shown (Kittel, 1976) that k_F, the maximum value of $|\mathbf{k}|$, is given by

$$k_F = \left(\frac{3\pi^2 N}{V}\right)^{1/3} \tag{3.3.}$$

and the Fermi energy itself by

$$E_F = \frac{\hbar^2}{2m_e}\left(\frac{3\pi^2 N}{V}\right)^{2/3} \tag{3.4}$$

where $V = L^3$ is the volume of the sample. The surface in $\mathbf{k}$-space connecting all wavevectors with energy E_F, known as the *Fermi surface*, has the shape of a sphere for free electrons. As we shall see later, for most real metals its shape is much more complicated. In addition to the actual value of E_F, other quantities commonly involved in a description of metallic phenomena are the velocity of electrons at the Fermi surface, v_F, which can be written as

$$v_F = \frac{\hbar k_F}{m_e} = \left(\frac{\hbar}{m_e}\right)\left(\frac{3\pi^2 N}{V}\right)^{1/3} \tag{3.5}$$

and the number of eigenstates per unit energy interval, often called simply the density of states, $D(E)$. For free electrons, this has the form

$$D(E) = \frac{3}{2}\frac{N}{E_F^{3/2}}E^{1/2} \tag{3.6}$$

The quantum free-electron model is able to account for the broad features of characteristic metallic phenomena such as electrical conductivity, thermal conductivity, heat capacity, magnetic susceptibility and thermionic emission. However, it breaks down badly in the face of detailed experimental observations, and is totally unable to explain the basic problem of why some solids are metals and others are not. The model also presents conceptual problems. It seems very surprising that the electrons can be regarded as non-interacting, since they possess electrical charge and collide frequently with each other. Furthermore, the assumption of a constant, zero, potential does not accord with the model of a solid as an infinite array of atoms or ions, each confronting an electron at close quarters with a potential very different from zero. The *one-electron model* assumes that the situation can be described by a Schrödinger equation for each individual electron

$$\left\{ -\frac{\hbar}{2m_e}\nabla^2 + V(\mathbf{r}) \right\}\psi = E\psi \qquad (3.7)$$

in which $V(\mathbf{r})$ is taken to contain, not only the electron-ion interactions, but also those between each electron and an average distribution of all the other free electrons. Bloch (1928) showed that, even without knowing anything more about the magnitude or form of $V(\mathbf{r})$, it is possible to write the solutions of (3.7) as

$$\psi_{\mathbf{k}}(\mathbf{r}) = u_{\mathbf{k}}(\mathbf{r})\exp i(\mathbf{k}\cdot\mathbf{r}) \qquad (3.8)$$

Here $u_{\mathbf{k}}(\mathbf{r})$ is a function which also possesses the translational symmetry of the lattice, and whose magnitude depends on the exact details of $V(\mathbf{r})$. The second factor, $\exp i(\mathbf{k}\cdot\mathbf{r})$, is just the travelling-wave solution for a free particle of wavevector $\mathbf{k}$.

Clearly, if $V(\mathbf{r})$ is a constant, say V_0, then $u_{\mathbf{k}}(\mathbf{r})$ is also a constant, and the free-electron picture results. It is reasonable, therefore, in seeking to model deviations from free electron behaviour, to expand $V(\mathbf{r})$ as a series and to investigate the difference that this makes to the wavefunctions $\psi_{\mathbf{k}}(\mathbf{r})$. Now, any function that has the same periodicity as the lattice can be written as a Fourier expansion in reciprocal lattice vectors, $\mathbf{g}$. For illustration, we limit the discussion to one dimension, although the arguments apply equally well to three dimensions. Then, with the integer $n \neq 0$,

$$V(x) = V_0 + \sum_{n=-\infty}^{\infty} V_n \exp i\left(\frac{2\pi n x}{a}\right) \qquad (3.9)$$

The fact that metals are predominantly free electron-like suggests that $V_n \ll V_0$. This is the nearly-free electron approximation. The mathematic details of the solution can be followed in Kittel, but we proceed directly to the conclusions. If the wavevector, k, corresponding to the wavefunction

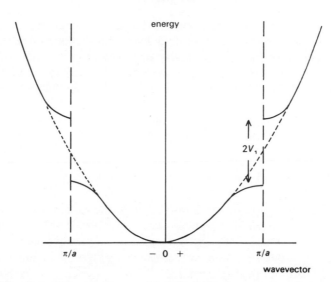

Figure 3.2 The effect of the periodic potential of the lattice is to open up a gap of forbidden energies at each Brillouin zone boundary. The quantity V_1 is the coefficient of the first term in the Fourier expansion of the periodic potential. The dashed curve indicates the free electron dispersion curve.

$u_k(x) \exp i(kx)$, is not close to a value $\pi n/a$, that is, to a Brillouin zone boundary (see §2.2), then the effect of the constant V_0 gives a constant $u_k(x)$ and the solutions are those for free electrons. If however, $k \simeq \pi n/a$, the term in the solution containing V_n becomes very large so that the periodicity of $u_k(x)$ dominates the wavefunction, and the electron no longer behaves as a free particle. An energy gap of forbidden frequencies opens up at $\pi n/a$, with a width equal to $2V_n$, as shown in Fig. 3.2.

An alternative view of the role of Brillouin zones is in terms of Bragg reflection of the free-electron waves at the appropriate zone boundary. Because of the reflection, the solution must be written as a standing wave. Two possibilities exist.

$$\psi(+) = \exp i\left(\frac{\pi n x}{a}\right) + \exp i\left(-\frac{\pi n x}{a}\right) = 2\cos\frac{\pi n x}{a}$$

$$\psi(-) = \exp i\left(\frac{\pi n x}{a}\right) - \exp i\left(-\frac{\pi n x}{a}\right) = 2i\sin\frac{\pi n x}{a}$$

(3.10)

It is clear that the two waves must have different energies. Whilst $\psi(+)$ has maxima centred on the ion cores and hence maximum interaction with them, $\psi(-)$ distributes most electronic charge in the regions between the ions, with minimum interaction. Whichever view is taken, therefore, the result is obtained that the electronic eigenstates may differ considerably from free-electron waves for wavevectors near Brillouin zone boundaries. Determination of the electron wavefunctions and dispersion curves in these regions is an important area of theoretical solid-state physics. The nearly-free electron model is only one approach, and the reader is recommended to look at Harrison (1981) for an introduction to tight binding, muffin tin potentials and pseudopotentials.

For low-temperature phenomena, the major concern is with the states and zone boundaries in the vicinity of the Fermi surface.

3.2 Studies of Fermi surfaces

We saw from the dispersion relation (3.2), that the surface of constant energy, E_F, for free electrons is a sphere. The only real metals that have even approximately spherical Fermi surfaces are the monovalent alkali metals, sodium, potassium, rubidium, and caesium. For these elements all the inner shells are filled, but the highest band, different of course for each element, is only half-full of electrons. The result is that the top of the electron distribution falls well short of the first Brillouin zone boundary and hence the effects of the periodic potential are minimal. Figure 3.3(a) shows the Fermi surface of potassium, circumscribed by the first Brillouin zone of its body-centred cubic structure. For comparison, Fig. 3.3(b) shows the Fermi surface of copper, which has a face-centred cubic structure. The electron distribution reaches further towards the zone face and in consequence suffers considerable distortion, but it is not difficult to see how Fig. 3.3(b) arises as a three-dimensional extrapolation of the one-dimensional dispersion relation, in Fig. 3.2. Things become more complicated still with divalent metals, since there are now sufficient electrons to spill over into the second Brillouin zone, as shown in Fig. 3.3(c) for beryllium, which has a hexagonal close-packed

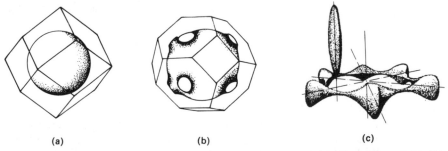

(a)   (b)   (c)

Figure 3.3   The Fermi surfaces of (a) potassium, body-centred cubic and monovalent: the Fermi surface is a free electron sphere to an accuracy of better than 1%; (b) copper, face-centred cubic and monovalent: the Fermi surface lies closer to the Brillouin zone boundary than for potassium and is distorted; (c) beryllium, hexagonal close-packed and divalent: electrons spill into the 2nd and 3rd Brillouin zones. The 'cigar' contains electrons, whilst the 'coronet' contains holes (Loucks and Cutler, 1964).

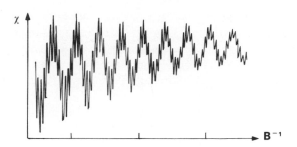

Figure 3.4 De Haas–van Alphen oscillations in zinc. The contributions from two separate orbits are clearly seen. (After Joseph and Gordon, 1962.)

structure. With the Fermi surface essentially intersecting the zone boundary, the effect of the periodic potential on the physical properties of the metal is now much greater; see Harrison (1981) and Shoenberg (1969).

Because the shape of the Fermi surface determines which electrons are available to take part in physical processes, a great deal of effort has gone into the experimental investigation of its characteristics for different metals. By far the most fruitful technique has been the de Haas–van Alphen effect, in which the magnetic susceptibility, χ, of the sample is measured as a function of an applied magnetic field, $\mathbf{B}$ (assumed to be the same as the internal field). Typical results are shown in Fig. 3.4. The magnetization of the electrons varies in an oscillatory fashion, and if χ is plotted against B^{-1}, where B denotes $|\mathbf{B}|$, it is found that the oscillations can be resolved into a small number of regular cycles. From the period of a cycle in terms of B^{-1}, the cross-sectional area of the Fermi surface in $\mathbf{k}$-space normal to the applied magnetic field can be inferred. Thus by varying the direction of the magnetic field the topology of the Fermi surface can be mapped out. The physics of the process is complicated and in the discussion we shall forsake mathematical rigour in favour of physical insight, beginning with a free-electron gas and extrapolating qualitatively to real metals.

A free electron in a magnetic field experiences a force

$$\mathbf{F} = e\mathbf{v} \wedge \mathbf{B} = \hbar \frac{d\mathbf{k}}{dt} \qquad (3.11)$$

Therefore a field applied parallel to z affects only k_x and k_y of the electron and, if k_z is equal to zero, in the absence of scattering the electron will describe a circular orbit in real space at the so-called cyclotron frequency, $\omega_c = eB/m_e$. What happens if k_z is not equal to zero? By substitution of $\mathbf{v} = \hbar\mathbf{k}/m_e$ in (3.11) we obtain the equation of motion of a two-dimensional harmonic oscillator with variables k_x and k_y that is completely decoupled from k_z. The energy eigenvalues of the total system are sums of the eigenvalues of k_z for a free particle, and of k_x and k_y for a two-dimensional harmonic oscillator. Thus

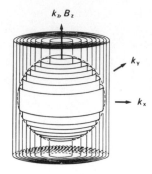

Figure 3.5 The quantization of the electron states in k-space due to a magnetic field giving rise to the Landau levels. The populated states lie on a series of cylinders within an approximately spherical envelope.

(Ziman, 1964)

$$E = \frac{\hbar k_z^2}{2m_e} + \left(n + \frac{1}{2} \right) \frac{e\hbar B}{m_e} \qquad (3.12)$$

where n is a quantum number taking integral values. The quantized states are called Landau levels, and all the states with the same value of n lie in **k**-space on a cylinder of cross-sectional area A, equal to $\pi(k_x^2 + k_y^2)$. Since the energy of the two-dimensional oscillator can be written as $\hbar^2(k_x^2 + k_y^2)/2m_e$, (3.12) can be regarded as a quantization of the area of the Fermi surface perpendicular to the applied magnetic field,

$$A = \left(n + \frac{1}{2} \right) \frac{2\pi eB}{\hbar} \qquad (3.13)$$

The extent to which the cylindrical shells are actually populated is determined by the Pauli exclusion principle, and instead of a spherical Fermi surface, the

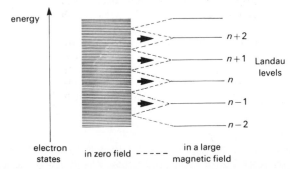

Figure 3.6 Showing how the quasi-continuum of states, all having the same k_z but different k_x and k_y, coalesce in the presence of a magnetic field to form the Landau levels. Since the total number of levels remains fixed, the number of states in each Landau level becomes increasingly large, while that between the Landau levels tends to zero.

shape shown in Fig. 3.5 results. Figure 3.6 presents the same information in rather a different way. The effect of the magnetic field is to 'bunch' together the electron states, which are distributed uniformly in the absence of the field, into the much narrower Landau levels. Since the total number of electron states must remain the same, the number of states contained in a single Landau level becomes very large, whilst forbidden gaps appear between the levels. The magnetic moment of the metal, and many other properties also, depend on $D(E_F)$, the density of states at the Fermi surface. If the magnetic field is varied, $D(E_F)$ will fluctuate between very small and very large values as the Landau levels are tuned through the Fermi level, the magnitude of which also varies as B^{-1}. From equation 3.13 it can be seen that, if B_1 and B_2 are the magnetic fields for two successive maxima of the oscillations in magnetic moment, differing in n by one, then

$$\frac{1}{B_2} - \frac{1}{B_1} = \frac{2\pi e}{\hbar A} \qquad (3.14)$$

It is clear that, for a three-dimensional Fermi surface, A is not uniquely defined by the magnetic field direction. There are very many electron orbits all perpendicular to $\mathbf{B}$. The ultimate subtlety of the de Haas–van Alphen effect and similar phenomena is that phase cancellation occurs for all orbits except those that are stationary with respect to $(k_x^2 + k_y^2)^{1/2}$, the so-called *extremal orbits*. Thus the maximum and minimum dimensions of the Fermi surface, which bound the extremal orbits, can be mapped out by varying the direction of the magnetic field. For a real metal with a complex Fermi surface, there might be several extremal orbits for a given direction of magnetic field (see Fig. 3.4). Different orbits correspond to different cyclotron frequencies, and indeed ω_c may be measured directly in many circumstances. The dependence of ω_c on the Fermi surface is often written in terms of a so-called *effective mass*, m^*, defined so that

$$\omega_c = \frac{eB}{m^*} \qquad (3.15)$$

where

$$m^* = \frac{\hbar^2}{2\pi}\left(\frac{\partial A}{\partial E}\right)_{E=E_F} \qquad (3.16)$$

The de Haas–van Alphen technique only works well at temperatures of a few kelvins, since the Landau levels are excessively broadened at higher temperatures. The thermal energy of the electrons themselves contributes $k_B T$ to the linewidth. In addition, scattering by impurities must be low enough to allow an electron to traverse its orbit several times, as otherwise lifetime broadening will also limit resolution of the Landau levels.

Measuring the magnetic moment is only one method of observing the Landau levels. The oscillatory behaviour of E_F and $D(E_F)$ shows up also in ultrasonic attenuation, when the electrons at the Fermi surface scatter acoustic phonons, in magnetoresistance, in thermoelectricity, and as mentioned above, in cyclotron resonance directly (Azbel and Kaner, 1956). The de Haas–van Alphen effect is preferred on the grounds of its high sensitivity, the ease with which the magnetic field direction can be changed, and the simplicity of the link between data and Fermi surface.

Of the many other properties that directly involve the electrons at the Fermi surface, heat capacity is of particular significance at low temperatures. The elucidation of the heat capacity due to electrons was another—we referred in chapter 2 to the phonon heat capacity—of the key successes of early quantum mechanics. Beginning with the assumption that the electrons are totally free, we suppose that the temperature of a metal is raised from absolute zero to T K. An electron can only acquire an amount of energy on average of about $k_B T$. But if $k_B T$ is much less than the Fermi energy, the only electrons that have empty states close by, and that are therefore able to be altered by the change in temperature, are those within $k_B T$ of the Fermi level. There are approximately $D(E_F)k_B T$ such electrons. The increase in internal energy of the whole electron system is therefore approximately $k_B^2 T^2 D(E_F)$, so that the heat capacity C_e, is $2k_B^2 T D(E_F)$. This is a very simple argument, but working out the calculation properly using the Fermi–Dirac distribution

$$f(E) = \frac{1}{e^{(E-E_F)/k_B T} + 1} \tag{3.17}$$

where $f(E)$ is the probability of a state of energy E actually being occupied, produces only a slight numerical change (Ashcroft and Mermin, 1976)

$$C_e = \tfrac{1}{3}\pi^2 k_B^2 T D(E_F) \tag{3.18}$$

Thus C_e gives a measure, albeit rather an insensitive one, of the electrons at the Fermi surface. For free electrons, we showed (equation 3.6) that $D(E_F) = 3N/2E_F$ whence C_e for free electrons, C_{fe}, comes out to be

$$C_{fe} = \frac{\pi^2 k_B^2 N T}{2E_F} \tag{3.19}$$

From this equation we can estimate that at room temperatures $C_e \sim 0.1\,\mathrm{JK}^{-1}\,\mathrm{mole}^{-1}$, compared with the phonon specific heat of $\sim 25\,\mathrm{JK}^{-1}\,\mathrm{mole}^{-1}$.

In fact it is only below a few kelvins for most metals that the electronic specific heat, because of its shallower temperature variation, begins to dominate. If C_e is written as γT, then the Sommerfeld constant γ has values for different metals varying between $2.22 \times 10^{-4}\,\mathrm{J\,mole}^{-1}\,\mathrm{K}^{-2}$ for beryllium and $180 \times 10^{-4}\,\mathrm{J\,mole}^{-1}\,\mathrm{K}^{-2}$ for manganese, reflecting the wide differences in the structure of the Fermi surface.

We have stressed that many processes involve specifically those electrons that have energies around E_F. In addition, the detailed shape of the Fermi surface determines various quantities that are needed in understanding the dynamics of electrons. We shall not give detailed derivations. Firstly, the actual velocity with which electrons travel in the solid is given by the group velocity, $\mathbf{v}_g$, of the electron waves. Thus (Ziman, 1964)

$$\mathbf{v}_g = \frac{\partial \omega}{\partial \mathbf{k}} = \frac{1}{\hbar} \frac{\partial E}{(\partial \mathbf{k})_{E=E_F}} \tag{3.20}$$

We have already mentioned the effective mass m^*, which can be written more generally as a tensor (Ziman, 1964)

$$m^* = \hbar^2 \left/ \left(\frac{\partial^2 E}{\partial \mathbf{k}^2} \right) \right. \tag{3.21}$$

and defines the response of the electron to an applied force. Finally, the density of states at the Fermi level is given in general by (Dugdale, 1977)

$$D(E_F) = \frac{V}{4\pi^3 \hbar} \oint \frac{d\Sigma}{|\mathbf{v}_g|} \tag{3.22}$$

where the integral is taken over the whole Fermi surface.

Information regarding these three quantities for a particular metal is a prerequisite for any attempt to understand its transport properties.

3.3 Transport properties of metals

Electrical conductivity, thermal conductivity and thermoelectricity are some of the most useful electronic properties of metals. (We shall discuss semi-conductors in §3.5.) The three phenomena are closely linked since they all arise out of the dynamics of a non-equilibrium distribution of electrons, and are commonly called *transport properties* for this reason. The non-equilibrium situation arises from the application either of an electric field, $\mathbf{E}$, or a temperature gradient, ∇T. It is convention to write the expressions for the electrical current density, $\mathbf{J}$, and the heat current density, U, in the symmetric form (Ashcroft and Mermin, 1976, page 254)

$$\begin{aligned} \mathbf{J} &= L_{11}\mathbf{E}' + L_{12}(-\nabla T) \\ \mathbf{U} &= L_{21}\mathbf{E}' + L_{22}(-\nabla T) \end{aligned} \tag{3.23}$$

where $\nabla V = \mathbf{E} + \nabla(E_F/e)$. From these equations it is easy to see how different transport properties may be selected by varying the conditions under which the experiments are carried out. For example, electrical conductivity is obtained by measuring $\mathbf{J}$ as a function of $\mathbf{E}$ at constant temperature, whilst the variation of $\mathbf{E}$ with ∇T with no current flowing yields the thermoelectric Seebeck effect. We begin by summarizing the major experimental features of the three phenomena.

The electrical conductivity, σ, of a metal and, similarly, its resistivity,

D

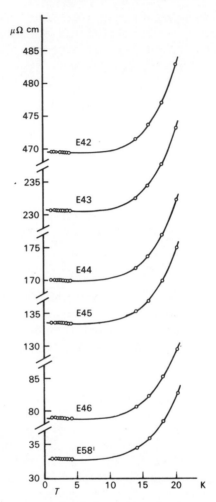

Figure 3.7 The low-temperature electrical resistivity of copper containing various concent-rations of tin in the parts-per-million range. The lowest concentration, 15 ppm, is at the bottom. (After Knook, 1962)

$\rho(=1/\sigma)$ at low temperatures are extremely sample-dependent, Figure 3.7 shows data for copper doped with various (low) concentrations of tin. The so-called Bloch T^5 law is observed in many metals at low temperatures. The residual resistivity at low temperatures is very sensitive to the presence of impurities and structural defects, and for a dirty specimen may not be very much smaller than that at room temperature. The *resistivity ratio*, which is defined by $\rho_{300}/\rho_{4.2}$, is a useful, rough measure of the purity of a crystal. Typical values for gold, silver, and copper obtained commercially are around 100, corresponding to a level of impurities of a few parts per million. This number can be substantially larger, as high as 10^5, in specially grown and zone-

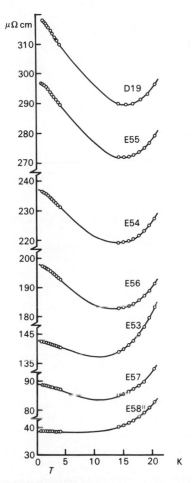

Figure 3.8 The resistivity of copper doped with iron in the parts-per-million range. The lowest concentration, 5 ppm, is at the bottom. In contrast to Fig. 3.7, a resistivity minimum occurs (after Knook, 1962).

refined samples. It is important to use specimens of high purity in de Haas–van Alphen and similar studies as the sharpness of the Landau levels is limited by the scattering by the impurities. Particularly dramatic effects on the transport properties are caused by magnetic impurities in metals. For example, in copper doped with only a few parts per million of iron, instead of settling down to a low temperature asymptote, the resistivity falls to a minimum at a few kelvins and then rises again as the temperature continues to decrease (Fig. 3.8). The *Kondo effect*, as it is called, is another manifestation of the basic exchange interaction between magnetic ions and conduction electrons, the RKKY interaction, that is also responsible for ferromagnetism in some metals (Kondo, 1964).

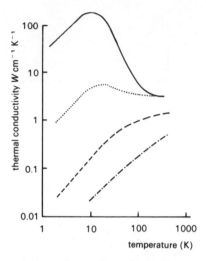

Figure 3.9 Thermal conductivity of some different forms of copper.
———— high-purity, annealed single crystal (Powell, Roder and Rogers, 1957);
.... typical industrial grade copper (*ibid.*);
– – – – typical sample of brass, (Cu 70:Zn 30) (Rosenberg, 1963);
–.–.–.– cupro-nickel (Cu 70:Ni 30) (White, 1979).

The thermal conductivity κ for most metals is dominated by the contribution from the electrons. Data for a few materials are shown in Fig. 3.9. The large peak present in pure materials is drastically reduced by the presence of impurities, a fact of great relevance in the construction of cryogenic equipment where a low heat influx is required. Since a flow of electrons is responsible for both electrical and thermal conduction in a metal, it is not surprising that a very direct relationship, the *Wiedemann–Franz law*, has been found to exist between these two quantities. For all metals at temperatures not too far below 300 K, the ratio $\kappa/\sigma T$ is approximately a constant, called the Lorenz number, L. We shall see that this result is easily understood on the basis of simple arguments about scattering, which predict that L_0, the value of L at room temperature, should be 2.45×10^{-8} W ohm K^{-1}. However, the Wiedemann–Franz law is not obeyed at all temperatures. Figure 3.10 shows that L

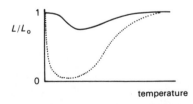

Figure 3.10 Dependence of the Lorenz number on temperature and sample purity. (schematic):
———— impure sample;
.... highest purity crystal attainable.

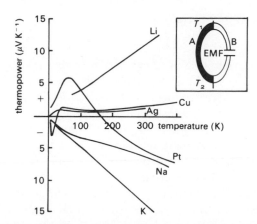

Figure 3.11 Temperature variation of the thermopower of some selected materials. Inset: schematic arrangement for measuring thermopower through the Seebeck effect.

eventually starts to fall as the temperature is decreased, but that at the lowest temperatures it rises again to its original value. The inference to be drawn from these data is that the mean free path of the electrons in electrical conduction is not always the same as that for thermal conduction. The significance of this result will become clear in §3.4.

The most familiar type of thermoelectric behaviour is the *Seebeck effect*, which forms the basis of the thermocouple, a useful thermometer for measuring low temperatures. When two dissimilar metals A and B are connected together, as shown in the inset of Fig. 3.11, with their two junctions at different temperatures and with no flow of electrical current, an EMF appears across the open circuit. Quantitatively related to this phenomenon is the *Peltier effect*, in which heat is evolved (or absorbed) at the junction between the different conductors as a current passes through the junction. In semiconductors, the effect can be rather large and can be used for macroscopic cooling. Finally, the *Thomson heat*, σ_T, is evolved (or absorbed) when a current flows through a conductor across the ends of which a temperature difference is maintained.

The quantity in terms of which all three effects are commonly expressed is the *thermopower*

$$S = \int_0^T \frac{\sigma_T}{T} \, dT \tag{3.24}$$

which relates to a single material. The similarity between this expression and (1.3) indicates why the same symbol is often used for both entropy and thermopower. The Seebeck EMF across the circuit in Fig. 3.11 is given by (Ziman, 1960)

$$\int_{T_1}^{T_2} (S_A - S_B) \, dT$$

and the Peltier coefficient π_{AB} by

$$\pi_{AB} = (S_A - S_B)T \tag{3.25}$$

Techniques for determining S are described by Barnard (1972). At low enough temperatures (below about 20 K) use can be made of the fact that the thermopower for a superconductor is zero, but above this temperature the measured Seebeck or Peltier effect is a differential property of two conductors. Figure 3.11 illustrates the thermopowers of some typical materials. The differences between similar materials are very striking, and in particular the fact that some thermopowers are positive and some negative in sign. Even more than for electrical conductivity, the data are extremely dependent on sample purity. Indeed, it is fair to say that thermoelectricity is so sensitive to fundamental electronic and lattice properties that it is not always a useful measure of them. We shall return to a discussion of some of the individual curves but firstly we wish to consider some common aspects of the three transport properties.

A derivation of expressions for σ, κ, and S from fundamental electronic properties using the Boltzmann transport equation would be out of place in this book. If the disturbance from an equilibrium distribution is small, if the relaxation time approximation is valid, and if electrons of each energy relax independently back to their equilibrium distribution, then the following expressions are obtained:

$$\sigma = L_{11} = \frac{e^2}{12\pi^3\hbar}\oint\tau(\mathbf{k})\mathbf{v}_g(\mathbf{k})d\Sigma \tag{3.26}$$

where $\tau(\mathbf{k})$ and $\mathbf{v}_g(\mathbf{k})$ are the relaxation time and group velocity for electrons of wavevector $\mathbf{k}$ and the integral is taken over the Fermi surface, Σ,

$$S = \frac{L_{21}}{L_{11}} = -\frac{\pi^2}{3e}(k_BT)^2\frac{1}{\sigma}\left(\frac{\partial\sigma}{\partial E}\right)_{E=E_F} \tag{3.27}$$

and

$$\kappa = L_{22} = \frac{\pi^2}{3e^2}k_B^2T\sigma \tag{3.28}$$

to an accuracy $(k_BT/E_F)^2$ and with the assumption that phonons contribute negligibly. Intermediate details can be obtained from Ashcroft and Mermin (1976), or in greater depth from Ziman (1960) and Mott and Jones (1936).

It is reassuring to find that 3.26 reduces to the Drude expression for free electrons and isotropic scattering. Then

$$\oint d\Sigma = 4\pi k_F^2,$$

and by use of (3.3), (3.26) becomes

$$\sigma = n_e e^2\tau/m_e \tag{3.29}$$

where $n_e = N/V$ is the number of electrons per unit volume. But the full expression (3.26) is necessary for real, anisotropic metals. If there are contributions to $\tau(\mathbf{k})$ from several different mechanisms which operate independently, then the total effect is obtained by adding reciprocal relaxation times, as for phonons in (2.26). Since $\tau^{-1}(\mathbf{k})$ is proportional to resistivity, ρ, this is equivalent to writing

$$\rho_{\text{Total}} = \rho_1 + \rho_2 + \cdots \qquad (3.30)$$

known as *Matthiessen's rule*. The close agreement with experiment, often to 1%, is surprising, in view of the condition that the different scattering mechanisms should provide independent channels for relaxation of the non-equilibrium distribution. The situation is discussed by Dugdale (1977).

Another experimental feature explained directly through equation 3.28 is the *Wiedemann–Franz law*. It should be noted that the validity of this result does not depend on an assumption of free electrons. The important condition is that the scattering should be predominantly elastic, and as we shall see below in §3.4, this is true both at low temperatures and at high temperatures. In the intermediate range the slight inelasticity of scattering by phonons becomes significant however, and the variations in the Lorenz number are due to this effect. In addition, there is a small contribution to the thermal conductivity from the phonons in a metal.

Thermopower curves such as those in Fig. 3.11 are more difficult to interpret, and (3.27) provides only a starting point. The most basic question is, why are some thermopowers positive in sign and some negative? The key here is the factor

$$\frac{1}{\sigma}\left(\frac{\partial \sigma}{\partial E}\right)_{E=E_F}$$

which it is helpful to rewrite as $(\partial \ln \sigma / \partial E)_{E=E_F}$. Since $\sigma = n_e e^2 \tau / m_e$, the sign of the thermopower depends on whether the number of electrons at the Fermi surface available for scattering, and τ, their relaxation time, increase or decrease as the Fermi surface expands. It turns out (see Ziman, 1960, for an explanation in greater depth) that the number of such electrons is very sensitive to the shape and size of the Fermi surface relative to the edges of the Brillouin zones. In particular, if the Fermi surface occupies more than about 50% of the first zone, its active surface (that area not in contact with the zone boundary) can actually decrease as the total volume increases, so that $(\partial \ln \sigma / \partial E)_{E=E_F}$ is negative. A quantitative calculation, however, is difficult and, in addition, the argument outlined above is clouded by other details. The large bump at around 60 K in the thermopower of platinum in Fig. 3.11 and the smaller ones for silver and copper are due to a second mechanism known as *phonon drag*. In this temperature regime the rate at which electrons can scatter from phonons, and hence transfer energy to a few well-defined lattice modes, can often be too rapid for the latter to be able to pass on the energy to the remainder of the phonon distribution. Thus an additional contribution to heat flow, and hence to the temperature gradient, arises from these phonons that are dragged along by the electrons. An illuminating description of the process has been given by Guénault (1971). Phonon drag makes only a tiny contribution to electrical conductivity but shows up strongly in thermopower. The effect is particularly large in semiconductors. It is difficult to make quantitative calculations because detailed models are required for both the electron and also the phonon scattering.

3.4 Scattering of electrons

The scattering of electrons in normal metals and semiconductors is dominated by the effects of impurities and lattice vibrations. It would seem that the electronic wavefunctions are especially sensitive to a break of the translational symmetry of the lattice, either static or dynamic. Interactions between the electrons themselves are much weaker, and there are few phenomena that show significant influence of electron–electron scattering. Kittel (1976) gives a qualitative explanation of this fact. In seeking to understand electron scattering processes, we note a valuable simplifying feature that is not present in phonon scattering. For most purposes, the

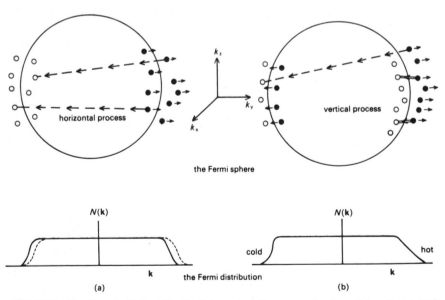

Figure 3.12   The disturbed Fermi distributions produced by the application of (a) an electric field to measure **J** and (b) a temperature gradient to determine **U**.
● excess of an electron relative to equilibrium distribution,
○ deficiency of an electron relative to equilibrium.
It is seen that electric field shifts the whole distribution in k-space, whereas temperature difference merely makes the distribution asymmetric.

electrons can be regarded as monoenergetic since many of them have energies within about $k_B T$ of E_F. We have seen also that E_F is typically a few eV whereas even at room temperature $k_B T$ is only $\sim 0.025\,\text{eV}$. Thus the scattering of electrons by thermal phonons is always quasi-elastic, and the magnitude of the phonon wavevector is the important quantity in a discussion of scattering. We find that the processes which cause electrical resistance are not necessarily the same as those responsible for thermal resistance.

The point can be understood by reference to Fig. 3.12 which shows schematically the non-equilibrium electron distributions produced (a) by an electric field and (b) by a temperature gradient. There is an important difference between the two distributions. The effect of an electric field is to shift the whole Fermi surface in **k**-space so that there is an excess of electrons with wavevectors in the positive direction and a deficit in the negative direction. Hence an electrical current flows. The effect of a temperature gradient is to produce an excess of positive hot electrons and an excess of negative cold electrons, but with no net shift of the distribution. In order for the distribution in Fig. 3.12(a) to relax, electrons must be scattered from $+ k_F$ to $- k_F$ across the Fermi surface, in so-called 'horizontal' processes, and phonons with wavevectors near the zone boundary are needed in order to accomplish this. In contrast, the thermal non-equilibrium can be relaxed even by small wavevector phonons in 'vertical' processes.

At room temperature, plenty of large wavevector phonons are available and the relaxation time for both electrical and thermal conductivity will be dominated by them. Thus the condition for the Wiedemann–Franz law is satisfied, that τ should be the same for **J** and **U**.

As the temperature is lowered, **J** will increase faster than **U** since the high wavevector phonons producing horizontal processes disappear more rapidly than the low wavevector ones which are responsible for vertical transitions. In the very low-temperature regime, the relaxation for both **J** and **U** is dominated by impurity scattering, which can reverse the direction of an electron in an elastic, horizontal process. Once again, therefore, the Wiedemann–Franz law is true. The reader is referred to Rosenberg (1963), page 115, for further explanation.

We shall now show how the extension of these ideas can explain some other features of the electronic transport properties of metals. We saw earlier that at low temperatures the electrical conductivity of many metals varies approximately as T^5. This temperature dependence can be accounted for through a fairly simple argument, by noting that the electrons are scattered only through small angles by phonons at low temperatures. It has already been noted that the collision of an electron with a phonon is quasi-elastic, that is, the electron is scattered to another point on the Fermi surface without change of energy. At low temperatures, k_F is much greater than the wavevector of the phonon **q** (which is approximately equal, at a temperature T, to $k_F(T/\theta_D)$) so that the maximum angle through which the electron can be scattered is of the order of

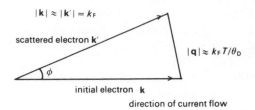

Figure 3.13 Illustration of small-angle scattering. The change in the component of the electron wavevector in the direction of current flow is $k_F(1 - \cos \phi)$.

T/θ_D. As illustrated in Fig. 3.13, the contribution ρ_{ep} made by this process to the resistivity is proportional to the decrease in forward component of $\mathbf{k}$; this is $k_F(1 - \cos \phi)$. The probability of scattering is proportional also to the square of the amplitude of the dominant lattice vibrations, which in turn varies as $k_B T$. Taking the integral over the Fermi surface from 0 to T/θ_D introduces an extra factor of $\sin \phi$. Thus

$$\rho_{ep} \propto T \int_0^{T/\theta_D} (1 - \cos \phi)\sin \phi \, d\phi) \tag{3.31}$$

For low temperatures, the scattering angle is small and the expression inside the integral reduces to ϕ^3. Finally, integration yields ρ_{ep} proportional to T^5. Low-temperature data for many metals fit the Bloch T^5 law reasonably well. At higher temperatures the picture is much less well-defined, and there is no universal analytical description of electron scattering by phonons. However, as a datum against which to compare experiment, the *Bloch–Grüneisen* expression is often used even though the model on which it is based is accepted to be unrealistic. Small-angle scattering is not assumed, but a spherical Fermi surface, elastic scattering and Debye phonons are built into the calculation. Ziman (1960), chapter 9, gives the mathematical details, with the result

$$\rho_{ep} \propto T^5 \int_0^{\theta_D/T} \frac{z^5}{(e^z - 1)(1 - e^{-z})} \, dz \tag{3.32}$$

The expression leads to the experimentally observed T and T^5 dependence at high and low temperatures respectively, but diverges from experiment at intermediate temperatures.

One further assumption made in the Bloch–Grüneisen model is that the electrons and phonons interact only through normal collisions and that there are no Umklapp processes. The nomenclature is exactly analogous to that for phonon scattering. Electrons at the Fermi surface of many metals lie quite close to a Brillouin zone boundary, so that a particularly energetic phonon can flip one of them into the next zone and cause a reversal of wavevector. The contribution of U-processes is difficult to identify experimentally since it has the same temperature dependence as the Bloch–Grüneisen expression at high and low temperatures. Ziman (1960) comments that U-processes may be responsible for 70% of the total room-temperature scattering in the alkali metals.

It is clear from Figs. 3.12 and 3.13 that small-angle scattering has a much greater effect in reducing the thermal current then it does in relaxing the electrical current. By contrast with merely slightly changing the direction of the electrical current, phonons can supply vertical processes which relax the hot electron distribution right back to thermal equilibrium. The temperature dependence of the contribution to the scattering simply follows the number of phonons, which goes as T^3. With an extra factor of T due to the specific heat of the electron gas, the temperature-dependence of the thermal conductivity at intermediate temperatures varies as T^{-2}, since $\kappa \propto C_V \tau$ for a free-electron gas, and τ varies as T^{-3}. At the lowest temperatures, where the electron scattering is due to impurities, τ is a constant, so that κ is simply proportional to T.

It is fortunate indeed that electrons are strongly scattered by impurities in metals. Without low-conductivity alloys such as stainless steel, cupro-nickel and constantan, the design and construction of cryogenic equipment would be very much more difficult and research at millikelvin temperatures would probably be impossible. Electrons are generally more strongly scattered by impurities than are phonons because, besides the break in translational symmetry, there is usually an excess or deficiency of charge at the defect. In addition, the wavelengths of electrons at the Fermi surface are often comparable with the physical size of the defect. The calculation of the residual resistance due to a particular impurity is usually an exercise in pure quantum mechanics with few intuitive simplifications. Comparison with experiment is made difficult by the lack of temperature dependence of the process. One useful relation that emerges is Nordheim's rule for the residual resistance, ρ_0, of an alloy consisting of two components in the ratio $x/(1-x)$. Then (Nordheim, 1931)

$$\rho_0 \propto x(1-x) \tag{3.33}$$

as long as basic parameters like electron density and shape of Fermi surface do not change with composition.

The simple first-order theory breaks down completely when the defect possesses a localized magnetic moment, such as iron or manganese (but not nickel) in copper. We noted earlier the large effects that small concentrations of these impurities can have on electrical conductivity and thermopower, as well as introducing an unexpected temperature dependence. For many years such data went unexplained, until Kondo (1964) solved the theoretical problem. He showed that the strength of the coupling between the spin of a conduction electron and that of an impurity moment is such that second-order terms in the scattering probability W_{ab}, can be as large as the first-order (a and b are the initial and final states respectively of the electron). Thus

$$W_{ab} = \frac{2\pi}{\hbar} \left(V_{ab}V_{ba} + \sum_{c \neq a} \frac{V_{ab}V_{bc}V_{ca}}{E_a - E_c} + \begin{array}{l} \text{complex} \\ \text{conjugate} \end{array} \right) \qquad (3.34)$$

where V_{ab}, V_{bc} and V_{ca} are the respective matrix elements. Pictorially, the incident electron is scattered first to an intermediate state, c, with a spin flip $(+)$, and is then scattered from c to b as a spin flip $(-)$ leaves the moment in its initial state. There is no simple argument beyond this point, but Dugdale (1977) provides a readable account of the steps leading to the expression

$$\rho = \rho_0 - \rho_K \ln T \qquad (3.35)$$

The suffix K refers, naturally, to Kondo. A more general review of localized moments in metals is given by Daybell and Steyert (1968).

3.5 Semiconductors and localization

There can be little doubt that, in technological terms, the most significant area of physics today is that of semiconductors. To an ever-increasing extent, industry and commerce are becoming dependent on the physical properties of doped single crystals of silicon. It is a truism perhaps that the research of the past decade, much of it carried out at low temperatures, has bred the technology of the present one. In this section, we shall introduce some of the new ideas and problems arising out of the research in the current decade.

First we must review briefly some of the basic properties of electrons in semiconductors and set them in the context of our earlier discussion in §2.1. The essential physical difference between metals and semiconductors lies in the relative positions of the Fermi level and the allowed energy bands. As we saw in §3.2, the Fermi surface of a metal lies at least partly within a band of allowed electronic states. In semiconductors, on the other hand, the electrons fill one of the bands right up to the top, so that at absolute zero there are no empty states close to E_F into which the electrons can be excited, as shown in Fig. 3.14(a). In terms of the Fermi–Dirac distribution (3.17), the situation is described as E_F having a value within a forbidden energy gap. At temperatures above absolute zero, electrons will be thermally excited from the top of the lower band (called the *valence* band) into the upper (*conduction*) band, in which they can move freely under the influence of, for example, an applied electric field—Fig. 3.14(b). The electrical conductivity will contain contributions both from these electrons, and also from the electrons remaining behind in the valence band, where there are now empty states into which they can move. It is convenient to describe the properties of the valence band in terms of a small number of positively charged holes rather than a large number of electrons. In contrast to metals, however, the total number of charge carriers is not constant but varies with temperature and other external parameters. The number of

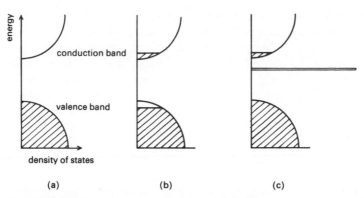

Figure 3.14 The density of electron states in semiconductors: (*a*) pure material at low temperatures ($\ll 300$ K) where few of the electrons in the valence band are thermally excited into the conduction band; (*b*) pure semiconductor at high temperatures (> 300 K); and (*c*) doped material at low temperatures. The impurity states, donors as shown in the figure, lie much closer to the conduction band edge and are thermally ionized at a lower temperature. (Hatched areas indicate populated states).

carriers excited also depends on the width, E_g, of the energy gap in which the Fermi level is situated. For silicon E_g has a value of 1.1 eV, but for other semiconductors values vary as widely as 0.2 eV (indium antimonide) and 4 eV (zinc sulphide). Any material with a higher-band gap would be regarded as an insulator because thermally excited carriers would be very few in number— but only if the sample was of high purity.

The presence of impurities influences drastically the properties of a semiconductor, and indeed many technological devices, such as integrated circuits, radiation sensors and injection lasers are dependent on carefully doped material. The primary effect of an impurity is to introduce into the crystal additional states whose wavefunctions are localized to the vicinity of the foreign atom and whose energies lie somewhere in the forbidden energy gap between the valence and the conduction band, as shown in Fig. 3.14(*c*). The best understood impurity systems are those which give energy levels lying close to the bottom of the conduction band, called *donors* since an electron occupying such a state can very easily be ionized into the band, or to the top of the valence band, known as *acceptors*. The wavefunctions for such impurities are approximately hydrogen-like, with the ionization energy scaled as $1/\varepsilon^2$ and with the Bohr radius, a_0, scaled by ε, where ε is the static dielectric constant of the material (Ashcroft and Mermin, 1976). However, many of the more interesting defects in semiconductors have energy levels lying deeper within the energy gap. Their properties cannot be explained on the basis of the simple hydrogenic model, although no generally applicable theory has yet emerged to replace it; see Jaros (1982).

Two basic techniques for characterizing semiconductors make use of the *Hall effect*, and electrical conductivity. The Hall effect is of particular value as a means of determining directly the number of charge carriers. If a steady magnetic field B_z is applied in a direction perpendicular to a flow of charged carriers having a velocity v_x in the x-direction, then the latter experience a transverse, Lorentz force, F_y, in the y-direction equal to $-ev_xB_z$. Initially, and very briefly, carriers will be deflected by F_y to the side of the sample, but an excess charge will rapidly build up producing an electric field E_y which cancels out the effect of the Lorentz force on later carriers. It is straightforward to show that

$$E_y = \frac{-j_xB_z}{ne} \tag{3.36}$$

where j_x is the current density and n is the number of carriers per unit volume. Thus a measurement of the Hall field, E_y, gives a direct determination not only of n, but also of the sign of the carriers, and hence whether they are electrons or holes. If both types are present, the expression for E_y is more complicated, and is given by Blakemore (1974), chapter 4. Furthermore, (3.36) is exactly correct only for free electrons and holes. A complete model would involve a detailed consideration of the shape of the Fermi surface. It may be noted that the magnitude of the Hall field varies inversely with n so that the effect is much larger for semiconductors than it is for metals.

The electrical conductivity, σ, of a semiconductor is usually expressed, in the presence of a single type of carrier, as

$$\sigma = ne\mu \tag{3.37}$$

where μ, the mobility of the sample, contains all the information regarding the scattering of the electrons or holes. At high temperatures (in the present context, this means up to 300 K) the major contribution to scattering of both electrons and holes is from the thermally excited phonons, giving a contribution to μ varying approximately as $T^{-3/2}$. At low temperatures, the carriers are scattered predominantly by defects, either neutral ($\mu \propto T^0$) or ionized ($\mu \propto T^{3/2}$) (Ziman, 1960). However, the main temperature dependence of σ arises from the variation of n. Figure 3.15 shows typical low temperature data for the resistance of germanium doped with approximately $10^{15}\,\text{cm}^{-3}$ of antimony which introduces donor states, plotted, following convention, as $\ln R$ against T^{-1} in order to display the exponential temperature dependences as straight lines. At the highest temperatures, all the impurities are ionized and the weak temperature variation arises solely from the mobility. Intrinsic conductivity occurs at too high a temperature to show up in these data. At temperatures below about 30 K the variation is due largely to thermal ionization of the donors. The fact that two separate straight lines with gradients in the ratio 2:1 are observed, with a transition between them at 4 K, indicates that the material is slightly compensated, containing about $10^{13}\,\text{cm}^{-3}$ of residual impurities possessing acceptor states, in addition to $10^{15}\,\text{cm}^{-3}$ of donors.

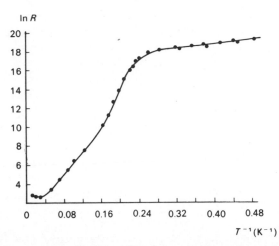

Figure 3.15 The variation of resistance with temperature for antimony-doped germanium containing $10^{15}\,\text{cm}^{-3}$ of donors and approximately $10^{13}\,\text{cm}^{-3}$ of acceptors.

In the absence of any other conduction process, the steeper slope would continue right down to absolute zero. However, as the number of thermally ionized donors and hence the number of free electrons in the conduction band continues to fall, so another process of charge transfer is revealed, that of thermally-assisted, quantum-mechanical tunnelling or *hopping*. Tunnelling is a direct consequence of the quantum mechanical nature of matter. Classically, a particle such as an electron would be viewed as being trapped on a particular site, $\mathbf{r}_1$, in the crystal, unable to escape because of the height of the potential barrier surrounding it. The quantum mechanical description of the impurity state is in terms of a wavefunction having a maximum at $\mathbf{r}_1$, but with a non-zero value throughout all space. There will always be a finite probability that an electron, initially in the eigenstate centred on $\mathbf{r}_1$, will appear at a later time in the state localized around a different site, $\mathbf{r}_2$. The model is illustrated in Fig. 3.16. The probability P_{12} of the electron tunnelling between the two states is related to the magnitude of the overlap, which falls off as

$$\exp\left(-2\frac{|\mathbf{r}_1 - \mathbf{r}_2|}{a_0}\right)$$

since at large distances each individual wavefunction decreases exponentially. P_{12} also depends on the difference $(E_1 - E_2)$ in energy between the states ψ_1, and ψ_2; this energy must be supplied, usually by thermal phonons, through the Boltzmann factor in order to conserve energy in the hopping process. Then, approximately (Mott and Davis, 1979),

$$P_{12} \propto \exp\left\{-2\frac{|\mathbf{r}_1 - \mathbf{r}_2|}{a_0} - \frac{E_1 - E_2}{k_B T}\right\} \qquad (3.38)$$

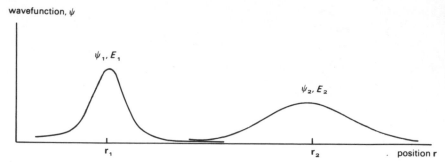

Figure 3.16 Illustrating the overlap between wavefunctions of different impurity states, that gives rise to hopping. Both eigenfunctions and eigenvalues are different so that energy must be supplied in order for hopping to take place.

The electrical conductivity of the crystal will be proportional to P_{12}, and the exponential relation predicted by (3.38) is indeed observed with many materials, as for example in Fig. 3.15 in the data below 4 K.

The type of hopping described above is known as *nearest-neighbour* hopping, because it takes place predominantly to the nearest available impurity sites, or *Miller–Abrahams* hopping after those who first articulated the concept (Miller and Abrahams, 1960). The condition for nearest-neighbour hopping to be the dominant mechanism is that $E_1 - E_2 \ll k_B T$. A conceptually different process occurs in heavily doped materials where the distances between the impurities are small so that $E_1 - E_2$ is large, or at very low temperatures. If the electron makes a transition to a state *lower* in energy so that $E_1 - E_2$ is negative, then it can hop to a more distant site for the same value of P_{12}. That is, it can jump further downhill than uphill! This process is called *variable-range hopping* and has a characteristic temperature dependence (Mott, 1969):

$$\sigma \propto \exp\left(\frac{A}{T^{\frac{1}{4}}}\right) \tag{3.39}$$

A physical explanation of this dependence is given by, amongst others, Apsley and Hughes (1974).

Hopping conduction at low temperatures is a topic of great current interest. We have introduced the ideas through the specific example of the conduction between impurities in semiconductors, but a major reason for the current research effort is the attempt to understand the electronic properties of non-crystalline materials and the more general phenomenon of *localization*. Quite apart from the intrinsic interest in such a fundamental, unsolved problem of long standing, it is certain that the semiconductor industry would save a great deal of money if single-crystal devices could be replaced by layers of glass. Many aspects of electrons in glasses are still only qualitatively understood, but it has become clear that, in general, the electronic eigenstates of truly

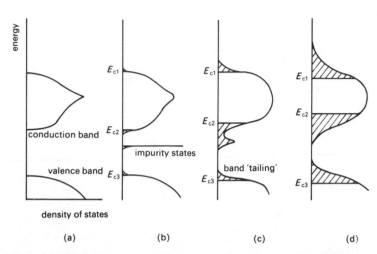

Figure 3.17 A speculative illustration of the effect of adding increasing numbers of structural and point defects to an initially pure crystal represented in (*a*) to create an amorphous material (*d*). Hatched areas indicate localized states, separated from the extended electron states by mobility edges, E_{c1}, E_{c2}, and E_{c3}. (After Economou *et al.*, 1974.)

amorphous materials, although grouped together in allowed bands as for periodic lattices, are predominantly not free electron waves at all but are *localized* states. It follows that the concept of a wavevector **k** is no longer a valid label for the electrons. Each wavefunction has a significant value only over a small region of space, not necessarily associated with any specific defect, and falls off exponentially away from this region.

The essential connection between randomness and localization, first shown by Anderson (1958), can most easily be visualized by realizing that a non-crystalline solid is no more than a crystal with a very large number of structural and other defects. The effects on the band structure of introducing an ever-increasing number of defects into a crystal are shown schematically in Fig. 3.17(*a*)–(*d*). For a perfect crystal (*a*) the electronic states have extended wavefunctions, so that an electron has equal probability of appearing anywhere within it. Each band has perfectly sharp edges with a density of states that varies as $E^{\frac{1}{2}}$. For Fig. 3.17(*b*) a small number of defects have been introduced. As we described above, impurities give rise to states in the energy gap which are localized around the defect atoms. Structural defects are also localized and the electronic states to which they give rise occur as a 'tail' on the edge of a free electron band. It can be proved that extended and localized states cannot co-exist at the same value of energy. Hence there must be a series of critical energies E_c, called *mobility edges*, that separate the extended from the localized energy levels.

As the concentration of defects is increased further, the band of localized impurity states broadens through interactions between the impurities, and

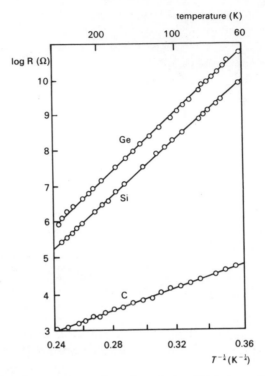

Figure 3.18 The resistance, R, of amorphous films of group IV elements, displaying variable-range hopping (Morgan and Walley, 1971).

may indeed overlap the advancing tail of the extended band. At the same time, the mobility edges move in towards the centre of the band of extended states (Fig. 3.17(c)). For a totally defective solid, Fig. 3.17(d), it is not at present certain whether a narrow band of extended states still remains or whether all the states in a truly amorphous material are localized. The relevance of hopping to these ideas is that it provides an experimental criterion for the presence of localized states. Since the bands of localized states are broad compared with $k_B T$ it is variable-range hopping that would be expected, and which in many samples is actually observed. Figure 3.18 shows data for amorphous films. Unfortunately in many samples the $\exp(A/T^{\frac{1}{4}})$ is not found when it might be expected. For this reason, theories continue to be tested on crystalline semiconductors doped with well-characterized impurities. In these systems, the randomness of the localized states can, it is hoped, be separated from the randomness of the structure itself.

Another fundamental topic in solid-state physics closely related to the present discussion is the *metal–insulator* transition. In view of the preceding paragraphs, it is natural to ask whether localized states can be changed into extended states merely by decreasing the distances between the centres of

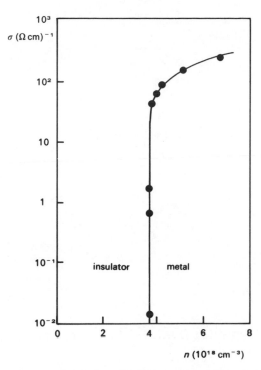

Figure 3.19 Low-temperature electrical conductivity, σ, of phosphorus-doped silicon of various phosphorus concentrations showing the metal-insulator transition (Rosenbaum *et al.*, 1980).

localization. Again the ideal system on which to make the test is the semiconductor–impurity system, in which the separation of the centres can be changed merely by varying the concentration. Figure 3.19 shows some striking data for various crystals of phosphorus-doped silicon obtained by Rosen-baum *et al.* (1980), suggesting that extended states can indeed be built up from localized ones, and that the transition is very sharp. Mott suggested that this sharp change from conduction by hopping to free-electron conduction, the so-called *Mott transition*, should occur universally at a defect concentration, n_c, given by

$$n_c^{1/3} a_0 = 0.26 \pm .05 \qquad (3.40)$$

The arguments by which this particular number is arrived at are explained by Mott and Davis (1968). Edwards and Sienko (1978) assembled data for a large number of materials, shown in Fig. 3.20, that support Mott's ideas.

A different model of a metal/non-metal transition was suggested by Anderson (1958) in terms of the bands of localized and extended electronic states shown in Fig. 3.18. He suggested that an *Anderson transition* occurs when, due to the variation of some external parameter, the Fermi level passes

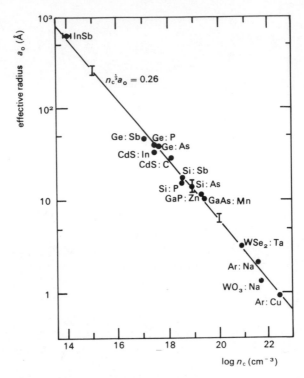

Figure 3.20 Comparison of Mott's criterion for the metal–insulator transition with data for various semiconducting systems (Edwards and Sienko, 1978).

through a mobility edge from a localized to an extended regime. (See also Mott *et al.*, 1975). So far, the most convincing evidence for mobility edges has been obtained, not in a three-dimensional solid, but for a two-dimensional electron system. The reader may wonder where in nature such an object might be found. There are, in fact, three situations in which its properties may be studied: in a very thin ($\sim$ 1 nm) metal film; in a monolayer of electrons trapped by their own image potential on the surface of liquid helium; and in a MOSFET (metal oxide semiconductor field effect transistor). This last involves some particularly elegant physics which makes an appropriate climax to our discussion of electrons at low temperatures.

The essential features of a MOSFET are that the electronic energy bands in the semiconductor are strongly distorted at the surface of the insulator, as shown in Fig. 3.21, and that the extent of the bending can be controlled by the application of an external 'gate' voltage, V_g. If V_g is made sufficiently large, then the bands are so bent that the Fermi level enters the conduction band and an inversion layer is created. The carriers in the inversion layer are electrons—in the device illustrated in Fig. 3.21—instead of the holes that are

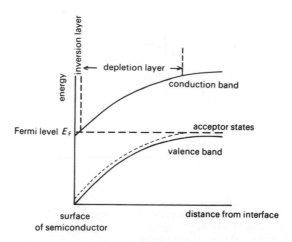

Figure 3.21   Potential in p-silicon at the interface with SiO_2 in a MOSFET (Mott *et al.*, 1975).

the majority carriers through the rest of the material. A fuller description is given by Pepper (1978). The inversion layer may be so narrow that at low temperatures only one value of k_z is allowed, that is, the layer is half a wavelength thick. Varying V_g over a wide range does not affect the two-dimensionality, but it does change the position of E_F. It has been found consistently that at a critical value of V_g the conductivity of the electrons in the layer alters in temperature dependence from $\exp(-B/T)$ to $\exp(-C/T^{\frac{1}{3}})$. The latter is the signature of variable-range hopping in two dimensions, and the interpretation of these observations is that as E_F was tuned through the mobility edge between extended and localized states, so the conduction mechanism changed from one of thermal ionization into one of hopping.

Another important effect that arises in the two-dimensional electron gas is the quantum Hall effect, first demonstrated by von Klitzing *et al.* (1980). If a large magnetic field is applied to the MOSFET, the electron distribution in the inversion layer, already only one state thick in k-space, is further quantized into rings in the k_x, k_y plane which are slices of the cylinders shown in Fig. 3.5. By varying V_g, the Fermi level can be positioned between Landau levels so that no electronic states exist at E_F. The situation is illustrated in Fig. 3.22. A measurement of the Hall effect under these conditions yields

$$E_y = \frac{j_x B}{N_L e} \tag{3.41}$$

where E_y is the Hall field, j_x the current density, B the magnetic field, and N_L is the total number of electrons occupying Landau levels inside E_F. But we can determine N_L also from a different principle. Since the application of the magnetic field does not alter the total number of states, it follows from (3.13) that the number of states per Landau level must equal the area of k-space

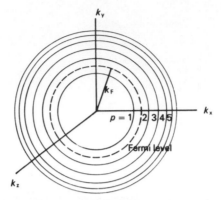

Figure 3.22 Magnetic field quantization in a two-dimensional electron gas. The Landau levels are only one electron state thick in the k_z direction. For illustration the Fermi level, E_F, has been drawn between the $p = 1$ and $p = 2$ Landau levels.

between successive Landau levels. If there are n Landau levels within E_F, then

$$N_L = \frac{neB}{h} \qquad (3.42)$$

Thus a careful measurement of E_y and j_x provides a determination of the quantity e^2/h which is essentially the fine-structure constant. Even in their preliminary experiments von Klitzing *et al.* were able to obtain an accuracy of a few parts per million. Since the time of the original experiments, many more interesting aspects of the phenomenon have been observed, implying that the two-dimensional electron gas is likely to be one of the key areas in semiconductor physics in the next decade.

Bibliography

References

Anderson, P.W. *Phys. Rev.* **109**, 1492 (1958).
Apsley, N. and Hughes, H.P. *Phil. Mag.* **30**, 963 (1974)
Azbel, M.I. and Kaner, E.A. *Sov. Phys. J.E.T.P.* **3**, 772 (1956).
Bloch, F. *Z. Physik* **52**, 555 (1928).
Economou, E.N., Cohen, M.H., Freed, K.F. and Kirkpatrick, E.S., in *Amorphous and Liquid Semiconductors*, ed. J. Tauc, Plenum, New York (1974).
Edwards, P.P. and Sienko, M.J. *Phys. Rev.* **B17**, 2575 (1978).
Guénault, A.M. *J. Phys. F* (Metal Physics) **1**, 373 (1971).
Jaros, M. *Deep Levels in Semiconductors*. Adam Hilger, London (1982).
Joseph, A.S. and Gordon, W.L. *Phys. Rev.* **126**, 489 (1962).
Knook, B., Ph.D. thesis, University of Leiden (unpublished, 1962): data in van den Berg, G.J., *Progress in Low Temperature Physics*, ed. C.J. Gorter, North Holland, Amsterdam, **4**, 194 (1964)
Kondo, J. *Prog. Theoret. Phys.* (Kyoto) **34**, 204 (1964).
Loucks, T.L. and Cutler, P.H. *Phys. Rev.* **A133**, 819 (1964).
Miller, A. and Abrahams, E. *Phys. Rev.* **120**, 745 (1960).

Morgan, M. and Walley, P.A. *Phil. Mag.* **23**, 661 (1971).
Mott, N.F. *Phil. Mag.* **19**, 835 (1969).
Mott, N.F. and Davis, E.A. *Phil. Mag.* **17**, 1269 (1968).
Mott, N.F., Pepper, M., Pollitt, S., Wallis, R.H. and Adkins, C.J. *Proc. Roy. Soc.* **A345**, 169 (1975).
Nordheim, L. *Ann. Physik* **9**, 607 (1931).
Powell, R.L., Roder, H.M. and Rogers, W.M. *J. Appl. Phys.* **28**, 1282 (1957).
Rosenbaum, T.F., Andres, K., Thomas, G.A. and Bhatt, R.N. *Phys. Rev. Lett.* **45**, 1723 (1980).
von Klitzing, K., Dorda, G., and Pepper. M. *Phys. Rev. Lett.* **45**, 494 (1980).

Further reading

Aschroft, N.W. and Mermin N.D. *Solid State Physics.* Holt, Rinehart and Winston, New York (1976).
Barnard, R.D. *Thermoelectricity in Metals and Alloys.* Taylor and Francis, London (1972).
Blakemore, J.S. *Solid State Physics.* W.B. Saunders, London (1974).
Daybell, M.D. and Steyert, W.A. 'Localised magnetic impurity states in metals', in *Rev. Mod. Phys.* **40**, 380 (1968).
Dugdale, S. *Electrical Properties of Metals and Alloys.* Edward Arnold, London (1977).
Engelman, R. *The Jahn-Teller Effect in Molecules and Crystals.* Wiley-Interscience, London (1972).
Harrison, W.A. *Electronic Structure and the Properties of Solids.* W.H. Freeman and Co., San Francisco (1981).
Kittel, C. *Introduction to Solid State Physics.* 5th edn., Wiley, New York (1976).
Melcher, R.L. 'The anomalous elastic properties of materials undergoing cooperative Jahn-Teller phase transitions', in *Physical Acoustics.* eds. W.P. Mason and R.N. Thurston, Academic Press, New York, **12**, 1 (1976).
Mott, N.F. and Davis, E.A. *Electronic Processes in Non-crystalline Materials.* 2nd edn., Clarendon Press, Oxford (1979).
Mott, N.F. and Jones, H. *Theory of the Properties of Metals and Alloys.* Clarendon Press, Oxford (1936).
Pepper, M., 'Two-dimensional systems' in *The Metal Non-metal Transition in Disordered Systems*, eds. Friedman and Tunstall, Scottish Universities Summer School Press, Edinburgh (1978).
Rosenberg, H.M. *Low Temperature Solid State Physics*, Clarendon Press, Oxford (1963).
Schoenberg, D., 'Electronic structure: the experimental results', in *The Physics of Metals— Electrons*, ed. J.M. Ziman, Cambridge University Press, Cambridge (1969).
White, G.K. *Experimental Techniques in Low Temperature Physics.* 3rd edn., Clarendon Press, Oxford (1979).
Ziman, J.M. *Electrons and Phonons*, Clarendon Press, Oxford (1960).
Ziman, J.M., *Principles of the Theory of Solids*, Cambridge University Press, Cambridge (1964).

4. Superconductivity

4.1 A low-temperature phase transition

Superconductivity occurs in some (but not all) metals, alloys and compounds. Above the transition temperature T_c, a metal behaves normally with no precursor of the dramatic changes in thermal and electrical properties which occur when the temperature is lowered below T_c. The most spectacular change is the total disappearance of the d.c. electrical resistance in the new phase. This property of superconductivity, first observed by Kamerlingh Onnes in 1911 when he used his liquid ^{4}He to cool mercury to 4 K, remains as yet an exclusively low-temperature effect and, although the highest transition temperature (T_c) has gradually increased over the years, first in alloys and then in binary and ternary compounds, room-temperature or even liquid-air-temperature superconductivity remains an as yet unattainable goal.

Current research activity is now directed towards materials which would not be classed as good electrical conductors at room temperature. There are a limited number of complex organic compounds with essentially one-dimensional structures that have the interesting and unusual property that they exhibit some of the characteristics of superconductivity at temperatures well above the point (usually about 1 K) at which a discontinuity occurs in the bulk resistance.

The link between room temperature conduction and T_c value is worth noting. It is those materials which would not be considered as good room-temperature conductors that have the highest transition temperatures: superconductivity has not yet been found in any of the pure noble metals; aluminium has a transition temperature of 1.2 K; while the intermetallic compound Nb_3Sn has a T_c value of 18.1 K. This correlation between high T_c and poor room-temperature conductivity provides an important clue to an understanding of the dominant mechanism for superconductivity.

The transition to the superconducting phase is strongly influenced by the presence of a static magnetic field (B_a) which, if it is sufficiently large, will cause a return to the normal phase. The combined effect of temperature and magnetic field is represented by the simple phase diagram shown in Fig. 1.9

Table 4.1 Transition temperature and critical fields for some of the more common elements

Element	$T_c(K)$	$\mu_0 H_c(0)(T)$
Al	1.12	0.011
Cd	0.52	0.030
Ga	1.08	0.051
In	3.41	0.028
Nb	9.25	0.206
Pb	7.20	0.080
Sn	3.72	0.031
Ta	4.47	0.083
V	5.40	0.141
Zn	0.85	0.053

(p. 18). The material will be superconducting for a combination of temperature and applied field, which lies below the solid line. Each superconducting element can be characterized by its transition temperature T_c and by its critical magnetic field* at zero temperature, $H_c(0)$, and Table 4.1 gives values of these quantities for some of the more common elements with $T_c > 0.1$ K. It will be noted that T_c and $H_c(0)$ scale together, although the exact correlation cannot be inferred from these data. Details for alloys and compounds will be given later in the chapter in the discussion of type II superconductivity.

In §4.2 we will see how the thermal and electrical properties of the superconducting state differ from those of metals in the normal state, and the existence of an energy gap between the superfluid ground state and the states of the normal electrons or quasiparticles will become apparent. Although quasi-particle tunnelling experiments actually postdate the microscopic theory of superconductivity, this simple but direct method of displaying the energy gap will nonetheless be described before the theory.

From time to time, we shall use some of the language and ideas of a *two-fluid model* for a superconductor. A similar concept is applied with considerable success to a description of the superfluid properties of HeII, and will be developed in chapter 5. The starting point is the twin postulates: first, that a superconductor or a superfluid possesses an ordered or condensed ground state which is characterized by an order parameter; and, secondly, that the entire entropy of the system resides in the excited states at energies above that of the condensate. Superconducting properties are associated with the 'superfluid' fraction (f) of the conduction electrons in the ground state, while the remaining 'normal' fraction ($1 - f$) retains the properties of electrons in the normal state at $T > T_c$. It should be emphasized that there can be no physical

---

* Some care will be needed when defining magnetic quantities. It is a convention that the symbol H_c is used to specify a critical field. However, we will want to refer to an applied field as B_a, measured in tesla, and so the critical fields are expressed in the same units.

separation of the two components, which are assumed to be totally interpenetrating and non-interacting. The main value of the model is that it
constitutes a simple physical picture providing a semi-quantitative understanding of, and the interrelation between, thermal and magnetic properties.

   Many of the thermodynamic and electrodynamic properties may be
explained or interpreted by phenomenological theories which do not require a
detailed knowledge of the ground state, or of the mechanism that causes the
transition to the condensed state. These will be discussed in §4.4 before the
rudiments of the remarkably successful theory of Bardeen, Cooper and
Schrieffer (hereafter BCS) are described. Because of their importance as the
source of a wealth of practical applications, some of which will be described in
chapter 8, the physics of the Josephson effect and the magnetic properties of
type II superconductors are treated in separate sections although their
relationship to the rest of the chapter will be emphasized.

4.2 Thermal properties

Unlike the λ-anomaly in ^{4}He (§5.1), the superconducting transition in zero
magnetic field is an almost perfect example of a second-order phase change,
without latent heat but with a sharp but finite discontinuity in the material's
heat capacity. Both the phonons and the conduction electrons contribute to
the heat capacity of a normal metal but there is little or no experimental
evidence to suggest that the lattice contribution changes in the superconducting state. The conclusion that the lattice properties are unaltered is supported
by other observations that the lattice constant, the crystal structure and the
elastic constants remain unchanged below T_c. Thus the sharp rise in the heat
capacity at the transition temperature indicates that the onset of ordering at T_c
is associated with a change in the entropy of the conduction electrons. If the

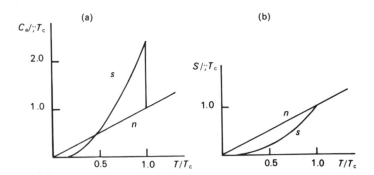

Figure 4.1   Temperature-dependence of the electronic heat capacity (*a*) and entropy (*b*) for a
normal metal and a superconductor.

lattice heat capacity is subtracted from the total, the temperature variation of the electronic component C_{es} will follow the line S in Fig. 4.1(a). The corresponding entropy of the conduction electrons decreases smoothly from its normal state value at T_c and is shown in Fig. 4.1(b). The more gradual behaviour of the entropy is consistent with a two-fluid model in which the proportion of electrons making up the superfluid fraction increases smoothly over a range of temperature, from zero at T_c to unity at $T = 0$.

A further result may be deduced from experimental heat capacity data if the values of C_{es} are plotted logarithmically against the inverse temperature. The results for gallium are shown in Fig. 4.2 and it will be seen that for $T < 0.6\,T_c$, the experimental points fit closely on the line whose equation is given by

$$\frac{C_{es}}{\gamma T_c} = A \exp - a\frac{T_c}{T},$$

where the normal-state electronic heat capacity is γT (§3.2) and a and A are constants. The exponential form of the equation is indicative of a finite energy gap between the ground state and the lowest excited states of the system. Only the normal fraction has any entropy and the number of thermally excited electrons is determined by the ratio (Δ/k_BT). The energy gap Δ is, in fact, temperature-dependent, falling from its zero temperature value $\Delta(0)$ to zero at T_c. For example, the value of $\Delta(0)$ for gallium derived from the data of Fig. 4.2 is 1.4 k_BT_c. It should be noted that the energy gap is much smaller than the Fermi energy k_BT_F, where T_F is typically 5×10^4 K. Other thermal properties are also affected by the superconducting transition. The low temperature thermal conductivity (κ) of a metal is dominated by the heat transported by the conduction electrons but, unlike its electrical counterpart, there is a decrease in the thermal conductivity of a pure material below T_c. This

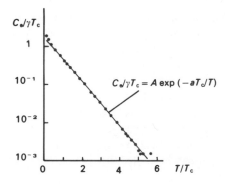

Figure 4.2 Electronic component of the heat capacity of gallium in the superconducting state. (After Phillips, 1964.)

observation is consistent, however, with the idea of a condensate of superelectrons with zero entropy. If these electrons are no longer able to exchange thermal energy with the lattice, there is no mechanism by which heat can be transported from one point to another. Thus it is the normal component that carries the heat current and it becomes decreasingly effective in doing so when it becomes very small at temperatures well below T_c.

The detailed temperature variation of κ differs from one material to another. Near T_c, these differences arise from the nature of the electron-scattering processes which may be caused by phonons or by impurities. At lower temperatures, when phonon conduction becomes dominant, the phonon mean free path is limited by electron scattering, and eventually by sample size. Because of the very large difference between the normal and superconducting state values of κ, superconductors make efficient low-temperature heat switches: for example the thermal conductivity of aluminium is reduced by a factor of 10^7 between T_c and $T_c/10$.

None of the conventional thermoelectric effects, thermopower, Peltier heat and Thomson heat occur in the superconducting state, since the electrical response of the normal electrons to a temperature gradient is effectively short-circuited by the superfluid condensate. However the thermoelectric response can be detected in a bimetallic loop made from two superconductors for, when the junctions are maintained at different temperatures, unquantized magnetic flux is generated within the loop. It will be shown in a later section that this effect is not possible in a loop of a single superconductor. Other thermoelectric effects which involve the movement of magnetic flux can be observed; and the reader is referred to a review by van Harligen (1981) for further details and references.

4.3 Electric, magnetic and electrodynamic properties

The change in the magnetic properties of a metal in the superconducting state is as dramatic as the disappearance of the electrical resistance. Although an applied magnetic field will eventually destroy the superconductivity, for all (or part) of the field range up to the critical value, the material is a perfect diamagnet: that is, the magnetic flux is expelled from the bulk of a superconductor irrespective of whether the field was first applied above or below the transition temperature. This Meissner effect (named after one of its discoverers) demonstrates that superconductivity involves much more than just a state of zero resistance, since that property by itself would cause flux to be trapped in a sample that was cooled through its transition temperature in a field less than its critical value.

The expulsion of flux from the bulk of a superconductor implies that, away from the surface region, the value of the average magnetic flux density or the B-field is everywhere zero. We write this result as $\langle B \rangle = 0$, where the average is over a scale of a few atomic spacings. This does not mean that the local field

$B(r)$ is zero everywhere; for it cannot fall from its external value to zero in a discontinuous manner at the surface of the sample. Rather, it has an exponential decay with a characteristic length λ, known as the penetration depth. $B(r)$ is related through one of Maxwell's equations to the local supercurrent density $J_s(r)$ by

$$\nabla \wedge \mathbf{B(r)} = \mu_0 \mathbf{J}_s(\mathbf{r}).$$

Two slightly different interpretations can be given to the Meissner effect. Flux exclusion occurs because the persistent supercurrents, which flow in the surface layer of thickness $\simeq \lambda$, generate a field which cancels exactly the applied field B_a so that, at a distance $\gg \lambda$ from the surface,

$$\langle B \rangle = M = 0,$$

where M is the sample magnetization. The field distribution outside the superconductor is then determined by B_a and the field due to surface current J_s. This is the first way of describing the effect; but the same distribution would be observed if the superconductor were replaced by a body of the same shape and size but with a finite magnetization. In the alternative description we may write

$$\langle B \rangle = 0 = \mu_0(H + M), \tag{4.1}$$

but with $$M \neq 0 \quad \text{and} \quad H \neq 0,$$
where M is the sample's average magnetization (i.e. total magnetic moment per unit volume) and H is the magnetizing field, which is generated by an external current, for example in a solenoid. Equation 4.1 then gives for the susceptibility χ,

$$\chi = M/H = -1,$$

which is the condition for perfect diamagnetism.

It is conventional to interpret the influence of the superconducting properties on the magnetic flux distribution in and around a superconductor in terms of its apparent bulk magnetization, rather than in terms of a distribution of surface currents $J_s(r)$. If we consider the case of a long thin cylindrical sample in a longitudinal magnetic field, the H field is simply B_a/μ_0 and the magnetization curve for an ideal superconductor would be as shown in Fig. 4.3. This behaviour is indeed exhibited by a large number of pure materials. At the critical field, H_c, the flux penetrates the superconductor and the normal state is restored. Any magnetic properties of the normal state are assumed to be too small to show up on the scale of this figure and, for $B_a > \mu_0 H_c$, $M = 0$. Magnetization curves of this type are not always observed, however, and two important exceptions will be discussed.

If the sample geometry is such that the field distribution around it is not homogeneous, the flux penetration will commence at an applied field less than the critical value. The penetration field H_p is determined by the demagnetizing

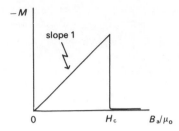

Figure 4.3 Ideal magnetization curve for a type I superconductor, which is in the form of a long thin cylinder. Between 0 and H_c, the material is in the *Meissner* state with $\chi = -1$.

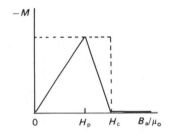

Figure 4.4 Magnetization curve for a spherical sample of a superconductor. Between H_p and H_c, the material is in the *intermediate state*.

factor for the particular sample geometry and for the limiting case of a very thin sheet in a transverse field, H_p goes to zero. The magnetization curve for a sphere is shown in Fig. 4.4. In the field region between H_p and H_c, the superconductor is in the *intermediate state* because the sample has broken up into regions which are alternately normal and superconducting, with the magnetic flux passing through the former. With some care, a simple planar laminar structure can be established which is of particular value for studies of the superconducting-normal interface. It should be emphasized that the occurrence of the intermediate state is a geometric effect and, for the remainder of this chapter, magnetic properties will be discussed in terms of samples with zero demagnetization factors in relation to the direction of the applied field.

The second deviation from the simple magnetization curve of Fig. 4.3 is a material effect which is illustrated by Fig. 4.5. As the magnetic field is increased, the superconductor first enters the Meissner state with the flux excluded from the bulk. At a field H_{c1}, called the lower critical field, flux penetration commences but the normal state is not fully restored to the bulk of the sample until the upper critical field H_{c2} is reached.* In the field region

---

* Superconducting effects in a surface sheath can be observed above H_{c2} but we will define H_{c2} as the maximum field in which a resistance less transport may be carried.

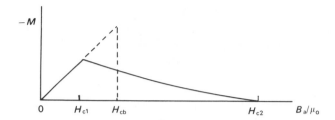

Figure 4.5 Magnetization curve for an ideal type II superconductor. Between H_{c1} and H_{c2}, the sample is in the *mixed state*, but it retains its superconducting electrical properties up to H_{c2}.

between H_{c1} and H_{c2} the *mixed state* structure differs from that in the intermediate state, in that the field penetrates the sample as isolated tubes of quantized flux rather than by forming normal regions (laminae) in which the field penetration is complete.

Superconductors which have a mixed state in their magnetization curves are classed as type II materials, while those which pass straight from the Meissner to the normal state are classed as type I. Pure samples of all but two of the elemental superconductors listed in Table 4.1—the exceptions are Nb and V—are type I: the alloys and compound superconductors, as well as the transition-metal elements which have a high resistance in the normal state, are type II. The detailed behaviour of type II materials is very sensitive to the metallurgical condition of the samples—a feature that will be returned to later in chapter 8—but their technological importance is simply illustrated by the values of H_{c2} given in Table 4.2. They are typically two orders of magnitude greater than the H_c values given in Table 4.1. By contrast, the lower critical field is usually much less than the thermodynamic critical field which is given the symbol H_{cb}. In Fig. 4.5 this field is defined by making the areas under the magnetization curve and the dashed line equal. This construction gives the type II material the same condensation energy as that of a type I material with $H_c = H_{cb}$.

Evidence that magnetic flux penetrates into the surface layer of a superconductor even when $B_a < \mu_0 H_c$ can be obtained from a simple experiment. When two coils are tightly wound on a superconducting core, the degree of electrical coupling between them is determined by the magnetic flux that passes through both of them. If there were total exclusion of flux from the superconductor as it was cooled below T_c, this coupling would disappear abruptly. The experiment shows that a small amount of coupling remains even at the lowest temperatures, and analysis of the results indicates that the penetration depth λ is temperature-dependent, decreasing from a value corresponding to total field penetration at T_c to an approximately constant value at $T_c/2$. Typically λ is about a few tens of nm for $T \ll T_c$.

Our discussion so far has centred on the response of a superconductor to a d.c. magnetic field when the induced surface currents are non-dissipative.

Table 4.2 Transition temperatures and upper critical fields H_{c2} for alloys and compounds

Material	$T_c(K)$	$\mu_0 H_{c2}(T = 4\,K)$
$Nb^{60}Ti^{40}$	9.5	13
Nb_3Sn	18.4	22
V_3Ga	15	23
$Nb_3(Al_{75} - Ge_{25})$	21	40

Measurement at a.c. frequencies above a few MHz shows that the surface impedance is complex. The reactive (inductive) component is related to the movement of flux within the penetration depth while the resistive component is a consequence of the motion of the normal fluid. The latter is a function of both temperature and frequency and, in the far infrared, absorption occurs even at $T = 0$. A typical set of experimental results is shown in Fig. 4.6. For radiation with photon energy $h\nu \ll k_B T_c$, a qualitative interpretation can be given in terms of the two-fluid model. The superconductor is replaced by an equivalent circuit made up of a pure inductance, representing the superfluid fraction, in parallel with a series combination of a resistance and an inductance, representing the normal electrons. As the temperature is lowered, the number of normal electrons falls, so that at $T = 0$, the parallel impedance would become infinite and the circuit would be loss-less.

The increase in the surface resistance R_s at higher frequencies would be explained if the number of normal electrons were being increased by photon excitation from the ground state. This is indeed the case and the threshold condition is given by

$$h\nu \geqslant 2\Delta(T) \qquad (4.2)$$

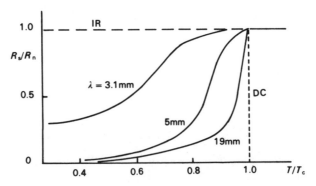

Figure 4.6 Temperature-dependence of the resistive component of the surface impedance of aluminium for different wavelengths of electromagnetic radiation λ_0. (After Biondi *et al.*, 1957.)

where $\Delta(0)$ is about the same as the energy gap found from the temperature dependence of heat-capacity measurements. The results of Fig. 4.7 confirm the temperature dependence of $\Delta(T)$ while the factor 2 in (4.2) suggests that the ground state involves pairs of electrons which can be separated by the absorption of a photon. Thus the results of two very different experiments have been interpreted successfully by introducing the concept of an energy gap. Measurement of the current-voltage $(I-V)$ characteristic of a tunnel junction provides further evidence for the existence of the gap and also a simple, direct way of finding its magnitude. This device is a sandwich of two metals which are separated by a thin insulating barrier. Particle tunnelling through a barrier is a familiar concept in quantum mechanics and its everyday importance can, of course, be seen by the passage of an electric current between two pieces of copper wire which are separated by a layer of grease or oxide. Giaever in 1960 first demonstrated how the $I-V$ characteristic of the junction is modified when either one or both of the metals became superconducting.

In the usual experimental arrangement shown in Fig. 4.8, two metal films

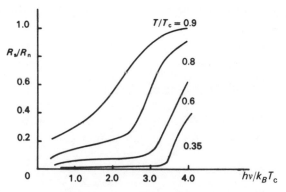

Figure 4.7 Resistive component of the surface impedance of aluminium as a function of radiation energy at different temperatures. (After Biondi and Garfunkel, 1959.)

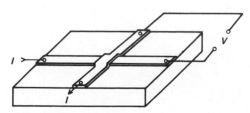

Figure 4.8 A cross-stripped tunnel junction made from evaporated films on an insulating substrate.

E

are evaporated as strips onto an insulating substrate (often a glass microscope slide) but the surface of the first film is oxidized before the second is evaporated over it. A typical junction area is 1 mm × 1 mm. The effects that are observed will depend on the thickness of the barrier: if this is less than about 2 nm and both metals are superconducting, a weak supercurrent will flow for zero voltage bias. This is called Josephson tunnelling, a topic which will be discussed in detail in §4.6. Electron or quasiparticle tunnelling is observed when the barrier thickness is between 5 and 10 nm.

When both metals in the junction are normal, the device shows ohmic behaviour. Since energy is conserved in the tunnelling process, which can only occur between full states on one side and empty states on the other, the total tunnelling probability is an integral which involves a somewhat complex product made up of Fermi factors and density of state functions for electrons at the Fermi level of each metal.

Figure 4.9 shows how the tunnelling characteristics of a symmetric SIS (superconductor-insulator-superconductor) junction change as the temperature is lowered below the common transition temperature of the two films. At $T = 0$, when there would be no normal electrons, the current would be zero up to a voltage V. The existence of an energy gap and the requirement that energy be conserved, combine to prevent any single-particle current flow until the bias voltage satisfies the condition

$$eV = 2\Delta(0)$$

At this point, current flow becomes possible as electrons from the ground state are excited into the normal electron states and there is a discontinuity in the characteristic. At $T > 0$, we note that the discontinuity becomes less pronounced and moves to a lower bias voltage, indicating once more the temperature-dependence of Δ, and also that current flow occurs at all voltages as the

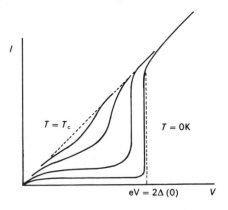

Figure 4.9 Typical I–V characteristics for a symmetric (SIS) tunnel junction at different temperatures.

number of thermally excited normal electrons increases. Later we will see that this sharpness of the discontinuity is a consequence of the form of the density of states function in a superconductor.

4.4 Theory of superconductivity

The first successful microscopic theory of superconductivity was given by BCS in 1957. The extent of its success can be gauged by its applicability to other systems (such as ^{3}He) and by the extent to which it can provide an explanation for almost all of the experimental properties of superconductors. Although the details of the theory are beyond the scope of this book, an understanding of the basic tenets of the model and of its predictions can be achieved without recourse to mathematical rigour. Many important experimental results can fortunately be described or explained without quantum mechanics or microscopic theory. Simple thermodynamic arguments can be applied to thermal properties, while the phenomenological equations of London augment Maxwell's equations to interpret the Meissner effect, to make predictions about flux quantization and to introduce the concept of the macroscopic wave function. The fact that the thermal and electrodynamic properties are given a more complete explanation by the BCS theory does not reduce the value of these other approaches, which prepared the ground for the success of the later BCS model.

Thermodynamics of superconductors

We have noted that the transition to the superconducting state is a function of temperature and applied magnetic field. In pure samples, the transition is reversible and can be described by equilibrium thermodynamics. The conditions for equilibrium are found by minimizing the magnetic Gibbs free energy at constant T and H. We will assume that there are no shape effects and therefore the magnetic field H will be applied field $(B_a/\mu_0) = H_a$.

Thus*

$$G(T, H) = U - TS - \mu_0 HM$$

Using the second law of thermodynamics,

$$dU = TdS + \mu_0 HdM,$$

where the second term is the external work done on the system. The condition for thermodynamic equilibrium is given by

$$dG = - SdT - \mu_0 MdH. \qquad (4.3)$$

---

* *Note*: The thermodynamic variables G, U and the extrinsic variables S, M are defined here for unit volume. In some texts, lower-case symbols are used for these quantities.

This equation can be integrated for an isothermal magnetization

$$G(T, H_a) - G(T, 0) = -\mu_0 \int_0^{H_a} M \, dH$$

In the normal state $(T > T_c)$, the magnetization is taken to be zero so that

$$G_n(H_a) = G_n(0)$$

In the superconducting phase $(T < T_c)$ and for $H_a < H_c$, $M = -H$, so that

$$G_s(T, H_a) = G_s(T, 0) + \tfrac{1}{2}\mu_0 H_a^2$$

The Gibbs free energy is increased by an amount equal to the magnetostatic energy of that volume of superconductor from which the applied field is excluded. The phase change $n \to s$ or $s \to n$ must occur when

$$G_s(T, H_c) = G_n(T, H_c)$$

This condition defines the thermodynamic critical field H_c for a type I superconductor, i.e.

$$G_n(T, 0) - G_s(T, 0) = 1/2\mu_0 H_c^2(T) \tag{4.4}$$

For applied fields greater than H_c, the normal state has the lower energy and is therefore stable. Equation 4.4 provides a direct measure of the condensation energy of the superconducting state, that is, the reduction in the free energy in forming the new phase is $1/2\mu_0 H_c^2(T)$ per unit volume of the sample. For a typical (type I) superconductor with $\mu_0 H_c = 0.1T$, this energy is $10^3 \, \mathrm{Jm}^{-3}$. Alternatively, expressed as an energy per conduction electron, it is $10^{-6} \, \mathrm{eV}$, a number which is very much less than the Fermi energy E_F and only about $10^{-3} k_B T_c$.

Equation 4.3 can also be used to determine the entropy from:

$$S = -(\partial G/\partial T)_H.$$

Since

$$G_n(T, H_a) - G_s(T, H_a) = 1/2\mu_0(H_c^2(T) - H_a^2)$$

then

$$S_n(T, H_a) - S_s(T, H_a) = -\mu_0 H_c(dH_c/dT)$$

Experimentally it is observed that dH_c/dT is always negative and that dH_c^2/dT is equal to zero only at $T = 0$ and $T = T_c$. Thus, although the entropy of the superconducting state is lower than that in the normal state, there is no discontinuity in S and hence no latent heat for the transition at T_c in zero field. For $T < T_c$, the transition to the normal state induced by a field involves a finite entropy change and, correspondingly, a latent heat. This is consequently a first-order phase change, whereas the zero field transition is second-order.

The discontinuity in the heat capacity at T_c then follows. Since $C = T(\partial S/\partial T)_H$,

$$C_{es} - C_{en} = +\mu_0 T_c (dH_c/dT)^2_{T = T_c}$$

Because the free energy in the two phases is equal at T_c, the critical field must go to zero at T_c.

From experimental observation, it is found that for all materials, the critical field H_c can be fitted, to within a few percent, to a simple square-law dependence on the temperature. It has the form (Fig. 1.8)

$$H_c(T) = H_c(0)[1 - (T/T_c)^2].$$

Electrodynamics of superconductors

A mathematical description of the response of a superconductor to an applied d.c. magnetic field must take account of both perfect conductivity ($\rho = 0$) and the Meissner effect or perfect diamagnetism ($\langle B \rangle = 0$). This was achieved by F. and H. London (1935) by two equations which, when used in conjunction with Maxwell's equations, predict that magnetic flux is excluded from all but a surface region of a bulk superconductor and that the decay of the field at the surface has a simple exponential form with a characteristic length. As we shall show below, one of these equations can be derived from London's postulate of a macroscopic wave function for a superfluid (§1.6).

The screening produced by an induced current in response to a changing magnetic field is a familiar effect, which is usually stated as Lenz's law. Although in a normal metal, the eddy currents are quickly reduced by scattering processes, screening currents on an atomic scale can persist and they give rise to a weak diamagnetic susceptibility in all materials.

In quantum mechanics, the current density $J(r)$ for charge carriers which are described by a single-particle wavefunction $\psi(r)$ is expressed by:

$$J(r) = -i\frac{e\hbar}{2m_e}[\psi^*\nabla\psi - \psi\nabla\psi^*] - \frac{e^2}{m_e}A(r)\psi^*\psi. \tag{4.5}$$

(This result is derived in chapter 21 of Feynman's *Lectures in Physics*, volume III).

Here $A(r)$ is the vector potential of any magnetic field that may be present and is related to it by

$$B(r) = \nabla \wedge A(r).$$

For steady-state conditions, we can make the assumption that the wavefunction $\psi(r)$ can be written as

$$\psi(r) = \psi_0 e^{iS(r)} = n_e^{1/2} e^{iS(r)} \tag{4.6}$$

where n_e, the number density, is equal to the probability density $\psi^*\psi$, and the phase $S(r)$ is a real function of position, and (4.5) becomes

$$\mathbf{J}(\mathbf{r}) = \frac{n_e e}{m_e} (\hbar \nabla S - e\mathbf{A}). \qquad (4.7)$$

If the wavefunction is unperturbed by an applied magnetic field, diamagnetic screening arises solely from the second form in (4.7). In a normal metal, the conduction electron wavefunctions are perturbed strongly by the field, so that the two terms almost cancel, to leave only a small residual diamagnetic response. This result suggests that the strong diamagnetism in a super-conductor is a consequence of the 'rigidity' of the electron wavefunctions. We will follow London's argument (§1.6) and assume that the condensate in a superconductor can be represented by a *macroscopic* wave function with the same form as (4.6). If we anticipate that the condensate is made up of pairs of electrons, number density n_s, mass $2m_e$ and charge $2e$, (4.7) becomes for a supercurrent density

$$\mathbf{J}_s = n_s \frac{e}{m_e} (\hbar \nabla S - 2e\mathbf{A}). \qquad (4.7a)$$

This equation can also be obtained from (1.21) by writing $J_s = n_s q v_s$ with $m = 2m_e$ and $q = 2e$. Setting $\nabla S = 0$, we obtain the first of the London equations

$$\mathbf{J}_s(\mathbf{r}) = \frac{-2n_s e^2}{m_e} \mathbf{A}(\mathbf{r}). \qquad (4.8)$$

The other equation is essentially a statement that, for an infinite conductivity, there can be no electric field inside a superconductor and equation 4.8 can be regarded as a magnetic Ohm's law insofar as the supercurrent J_s is driven by a magnetic field. Flux exclusion follows from (4.8) and Maxwell's equation for static fields:

$$\nabla \wedge \mathbf{B} = \mu_0 \mathbf{J}_s$$

Combining these two equations, we find

$$-\nabla \wedge \nabla \wedge \mathbf{B} = \nabla^2 \mathbf{B} = \mathbf{B}/\lambda_L^2 \qquad (4.9)$$

where

$$\lambda_L^2 = m_e/2n_s e^2 \mu_0$$

The solution of equation 4.9, for a magnetic field B_a just outside and parallel to the plane surface of a superconductor, is

$$B(x) = B_a \exp(-x/\lambda_L)$$

where x is the distance measured from the surface. The distance λ_L is known as the London penetration depth. If we take n_s to be half of the total density of

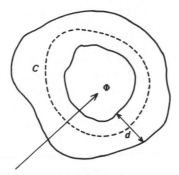

Figure 4.10   Flux quantization for a ring of superconductor with dimensions $d \gg \lambda_L$.

conduction electrons in the metal, the value of λ_L for tin is about 35 nm. However if n_s is only the fraction of electrons in the condensate, λ_L will be temperature-dependent, and according to the two-fluid model, it should vary as

$$[1 - (T/T_c)^4]^{-1/2}$$

Measurements of the penetration depth generally show good agreement with this form, but in most of the superconducting elements, the magnitude of λ extrapolated to $T = 0$ is greater than that expected on the basis of (4.9).

The more general form of the London equation (4.7a) predicts an interesting result when the superconductor has a hole passing through it, as shown in Fig. 4.10. Unlike a solid block in which the phase of the wavefunction must be single-valued, the presence of a hole relaxes the constraint that the change in phase around a closed path must be zero. If the dimension of the super-conductor is much greater than λ_L, we can make a contour integral around the dotted path so that J_s is everywhere zero. Thus

$$\oint \mathbf{J}_s \cdot d\mathbf{l} = 0 = \frac{n_s e}{m_e}\left[\oint \hbar \nabla S \cdot d\mathbf{l} - \oint 2e\mathbf{A} \cdot d\mathbf{l} \right]$$

Since the phase S can be increased or decreased by multiples of 2π without altering ψ, we find that

$$2\pi n = \frac{2e}{\hbar}\int (\nabla \wedge \mathbf{A}) \cdot d\mathbf{S} = \frac{2e}{\hbar}\int \mathbf{B} \cdot d\mathbf{S} = \frac{2e}{\hbar} \Phi \tag{4.10}$$

where we have used Stokes' theorem to change a line integral to a surface integral, and where Φ is the magnetic flux which passes through the hole. Although flux is excluded from the bulk of the superconductor, (4.10) tells us that flux can pass through a hole in it, but only in units of $h/2e$. The flux quantum is extremely small, having a value of 2.10^{-15} Wb. Experimental confirmation of this quantization was provided by Deaver and Fairbank

(1961) by measuring the magnetic moment of a very small superconducting cylinder, made by depositing a layer of tin onto a short length of (normal) copper wire of diameter 10^{-2} mm.

Perhaps the most important result to come from the work of the Londons is the recognition that a superconductor is characterized by a macroscopic wavefunction. Equation 4.5 for the current density is based on a single-particle wavefunction of the form $\psi_0 \exp(iS)$. If the London equation is to be applicable to a many-electron system, the phase of all the particles (pairs, as we shall later confirm) must be the same so that the total ground-state wavefunction will be made up of the product of a very large number of identical-pair wavefunctions. The exclusion principle would prevent this happening to single electrons, but this constraint is removed for pairs of electrons which behave as bosons.

The London penetration depth can be thought of as a screening length for magnetic fields (and currents) in a superconductor or as the minimum distance over which the field may be significantly changed. Pippard (1953) suggested that there was a second characteristic length which provided a measure of the minimum distance over which the ground-state wavefunction could be changed. This intrinsic coherence length ξ_0 can be estimated from energy uncertainty arguments. A special variation in ψ involves additional kinetic energy. The spatial variation must therefore be restricted such that the additional energy is less than the condensation energy of the superconductor. This argument gives ξ_0 as

$$\xi_0 = ahv_F/k_B T_c,$$

where v_F is the Fermi velocity and a is a numerical constant of order unity. The value of ξ_0 for tin is 230 nm and for many metals $\xi_0 > \lambda_L$. Pippard realized that this result would influence the electromagnetic response and in particular that the range of validity of the London equation is limited. The problem is similar to that of the high-frequency response of a normal metal when the electron mean free path is greater than the skin depth δ_n (or the minimum spatial variation of the electric field). The normal skin effect is based on a local relationship between the current density at a point r and the electric field at the same point. This is valid for $\delta_n > l$, the electron mean free path. Likewise the London equation is a local equation relating the local supercurrent density to the local vector potential, with a range of validity given by

$$\xi_0 < \lambda_L,$$

the spatial variation of $A(r)$. Non-local electrodynamics which applies in a normal metal when $l > \delta_n$, predicts a new screening length (the anomalous skin depth) which is proportional to $(l\delta_n^2)^{1/3}$. This suggests that in a super-conductor, the appropriate screening length in the situation when $\xi_0 > \lambda_L$ should be $\lambda \propto (\xi_0 \lambda_L^2)^{1/3}$, or a screening length which is greater than λ_L.

Pippard proposed that London's local equation should be replaced by an

Table 4.3 Values of λ and ξ for Type I superconductors

Element	λ_L(nm)	ξ_0(nm)	λ_{theory}(nm)	λ_{exp}(nm)	λ_{exp}/ξ_0
Al	16	1600	48	50	0.03
Cd	110	760	136	130	0.17
Nb	39	38	(24)	44	1.1
Pb	37	83	32	39	0.45
Sn	35	230	48	51	0.22

equation for $\mathbf{J(r)}$ which provides a suitably weighted average of $\mathbf{A(r')}$ over a sphere of radius ξ about the point $\mathbf{r}$. The value of ξ is not quite the same as ξ_0, since the true coherence length is modified by scattering, such that

$$\frac{1}{\xi} = \frac{1}{\xi_0} + \frac{1}{l} \tag{4.11}$$

Using the non-local form for $\mathbf{J(r)}$, Pippard was able to calculate the penetration depths for various values of ξ_0 and λ_L and make comparisons with experimental data. Results for both aluminium and tin could be fitted by the choice of a single parameter $a = 0.15$ in the expression for ξ_0.

Superconductors are sometimes classed as Pippard or London depending on the relative sizes of λ and ξ. The London superconductors for which local electrodynamics apply have

$$\lambda > \xi \quad \text{and} \quad \lambda(0) = \lambda_L \quad \text{if} \quad l \gg \xi_0$$

Pippard superconductors are those for which $\lambda < \xi$, $\lambda > \lambda_L$ and non-local electrodynamics must be used.*

Table 4.3 provides a comparison of the measured and calculated values of the parameters that determine the electrodynamic response of various superconductors. The measured values of both λ and ξ are functions of the mean free path; ξ decreases and λ increases with decreasing l. Thus by making an alloy with a short mean free path, it is possible to convert a Pippard superconductor into a London superconductor. This property will be seen to have important consequences in the discussion of type II superconductivity.

BCS theory

The unique quality of the BCS theory lies in its ability to provide an understanding and an explanation of all of the properties so far encountered. It

---

* It should be noted that the penetration depth λ is a function of temperature while the coherence length as defined by equation 4.11 is not. Another quantity, also called the coherence length, is used in the Ginzburg–Landau description of the mixed state of a type II superconductor (§4.7) and it is a characteristic length for the spatial variation of the wavefunction, $\psi(r)$. This quantity has the same temperature-dependence as λ and diverges at T_c. The difference between the Pippard and the Ginzburg–Landau coherence lengths is not important, except near T_c.

is a microscopic theory, based on a simple model of the dynamics of conduction electrons in a metal, which leads to quantitatively correct results and predictions for almost all superconductors. The fact that the model should be based on a spherical Fermi surface is at first sight somewhat surprising but it is worth noting that the existence of superconductivity does not depend strongly on crystal structure which does have a direct influence on the Fermi surface. Fundamental to the BCS theory is the hypothesis that under certain conditions there can be an attractive interaction between electrons which involves the positive ion lattice. Although we have noted that the lattice properties such as the heat capacity and the lattice parameter are not influenced by the transition, the indirect involvement of the lattice can be inferred from the isotope effect. Experiment shows that for an individual element the transition temperature T_c depends on the isotopic mass of the constituent atoms as

$$T_c \propto M^{-1/2}$$

The fact that metals with the highest room-temperature resistances and hence the strongest electron–phonon coupling have the highest values of T_c provides further evidence of lattice involvement. Fröhlich (1950) was the first to show that an interaction between conduction electrons and acoustic lattice vibrations could lead to an attractive inter-electronic force and that the filled Fermi sea was not necessarily the lowest energy state of a metal. It is something of a surprise that while the normal-state conductivity may be limited by electron scattering by thermal phonons, a phonon-induced electron–electron interaction leads to an ordered ground state with the property of infinite conductivity.

A convenient if incomplete picture of the attraction between electrons is as follows. An electron moving through the positively charged lattice will cause a perturbation or movement of the ions from their equilibrium positions. This in turn leads to a slight increase in the local positive charge density and a second electron will be attracted by this charge distortion. Even though the attractive force is weak, it may be stronger than the Coulomb repulsion so that the net force is attractive. An alternative pictorial representation of the electron–electron interaction is shown in Fig. 4.11 as a scattering process. An electron occupying a plane wave state $\mathbf{k}_i$ (i.e. a Bloch state $\sim \exp(\mathbf{k}_i \cdot \mathbf{r})$) emits a phonon with wavevector $\mathbf{q}$ which is subsequently absorbed by a second electron occupying the state $\mathbf{k}_j$. The total momentum is conserved in the process but it is not necessary to conserve energy as the scattering takes place in a very short time, and the energy is only defined within the limits of the uncertainty principle. This scattering diagram is also appropriate to the Coulomb interaction except that in this case the exchanged particle would be a photon.*

---

* An attractive interaction between electrons (fermions) is an essential component of the BCS theory. However the theory is not sensitive to the nature of the interaction and the pairing mechanism described here is not necessarily appropriate to all superconductors.

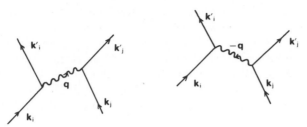

Figure 4.11   Electron–electron coupling via phonon exchange. In a normal process $\mathbf{k}_i' = \mathbf{k}_i + \mathbf{q}$.

In terms of second-order perturbation theory, the scattering event is described by a transition probability. A second process in which the phonon, wavevector $-\mathbf{q}$, is emitted by the electron in $\mathbf{k}_j$ and later absorbed by that in $\mathbf{k}_i$, leads to the same final state but the transition probability includes slightly different energy terms. When the two processes are combined, the strength of the electron–electron coupling is expressed as a potential $V(k_i, k_j)$ and for a small range of energies

$$|E(k_i) - E(k_j)| \leqslant \hbar\omega_q \tag{4.12}$$

it is negative and the force is attractive. This was the gist of Fröhlich's argument in 1950 but it was not until 1956 that Cooper was able to show that the formation of paired states could lead to a reduction in the total energy of the conduction electrons.

In the normal state of a metal at $T = 0$, the conduction electrons are assumed to be non-interacting, with all energy states up to a maximum of E_F being occupied in accordance with Fermi–Dirac statistics. These electrons occupy states in **k**-space within a sphere of radius k_F. Cooper considered what would happen if two additional electrons were added to a metal at $T = 0$. As all states up to k_F are filled, these electrons would have to occupy states with $k > k_F$, but he was able to show that if there is an attractive force between these extra electrons, the total energy of the bound state is less than $2E_F$. In other words, however small the interaction energy V might be, the increased kinetic energy of two electrons in states with $k > k_F$ is offset by the reduction in the potential energy of the bound state. A number of features in Cooper's analysis, which are relevant to the BCS theory, should be noted. As a result of the process described by Fig. 4.11, the pair of electrons will be scattered repeatedly by the emission and re-absorption of a phonon in such a way that the sum of the individual momenta is conserved, i.e.

$$\mathbf{k}_i + \mathbf{k}_j = \mathbf{K} \tag{4.13}$$

In quantum mechanical language, the pair wave function ψ_P is a sum of terms, each of which is the product of two Bloch functions $\psi(\mathbf{k}_i)$ and $\psi(\mathbf{k}_j)$, multiplied by an amplitude a_{ij} which is related to the probability of finding the electrons in the states $\mathbf{k}_i$ and $\mathbf{k}_j$. Cooper made the assumption that the scattering amplitude for each event is a negative constant, independent of $\mathbf{k}_i$ and $\mathbf{k}_j$. This attractive potential holds for electrons with almost the same energy, as required by 4.12, but both values of $|\mathbf{k}_i|$ and $|\mathbf{k}_j|$ must be greater than k_F. These conditions may be satisfied if the energies of the two electrons lie within a band above the Fermi level E_F equal to the average phonon energy, i.e. about $\hbar\omega_D/2$ where ω_D is the Debye frequency. This result is usually expressed as a range of momentum (Δk), measured with respect to k_F, within which $|\mathbf{k}_i|$ and $|\mathbf{k}_j|$ must lie, i.e.

$$V(\mathbf{k}_i, \mathbf{k}_j) = -V \quad \text{for } \Delta k \simeq m_e\omega_D/2\hbar k_F$$
$$= 0 \text{ otherwise.} \tag{4.14}$$

A simple geometric construction (see Fig. 4.12) can be used to determine which values of $\mathbf{k}_i$ and $\mathbf{k}_j$ satisfy the conditions imposed by (4.13) and (4.14). For each value of $\mathbf{K}$, the number of pair states that can take part in the attractive interaction is proportional to the volume swept out when the shaded area is rotated about $\mathbf{K}$. This volume has a sharp maximum when $\mathbf{K} = 0$ which is equal to the volume of a shell with thickness Δk. Thus the maximum reduction in potential energy is achieved when electrons are paired with equal and opposite momenta and it may also be shown that V is largest when the electrons have opposite spins. This quasi-molecule, called a Cooper pair, has a binding energy $2\hbar\omega_D\exp(-2/D(E_F)V)$ in the limit of $D(E_F)V \ll 1$ where $D(E_F)$ is the density of states for a single spin at the Fermi surface.

The extension of the Cooper model for a single pair of electrons to 10^{29} m^{-3} electrons in a real metal was the achievement of BCS. A basic assumption of their theory is that the only interactions which need to be considered are those between any two electrons that form a Cooper pair. The influence of all other electrons on this pair is to reduce the number of states into which the interacting pair would otherwise scatter since they may be occupied. Starting with a full Fermi sea at $T = 0$, two electrons with momenta just less than k_F are taken outside the Fermi sphere to form a Cooper pair occupying the states $(\mathbf{k}, -\mathbf{k})$, such that there is a reduction in overall energy. After the first pair has

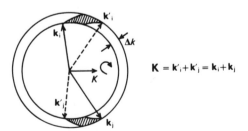

$$\mathbf{K} = \mathbf{k}'_i + \mathbf{k}'_j = \mathbf{k}_i + \mathbf{k}_j$$

Figure 4.12   Electrons in states $(\mathbf{k}_i, \mathbf{k}_j)$ are scattered into new states $(\mathbf{k}'_i, \mathbf{k}'_j)$ lying within a shell of thickness Δk from the Fermi momentum k_F, such that the total momentum is conserved. The shaded region, when rotated about $\mathbf{K}$, gives the interaction volume. It has a very sharp maximum when $\mathbf{K} = 0$.

been formed, the only restriction on the second is that the available states that can be occupied are now limited by the Pauli exclusion principle. We cannot continue to gain (negative) potential energy by removing electrons from below E_F and forming pairs with $k > k_F$ and the limit is reached when the gain in kinetic energy is no longer compensated by the reduction in potential energy. The condition which gives the lowest overall energy is expressed through a quantity v_k^2, which is the probability that the pair state $(\mathbf{k}, -\mathbf{k})$ is occupied in the ground state. The ground-state wavefunction in which all the n_e electrons are paired, is the product of $n_e/2 = n_s$ individual-pair wavefunctions of the type described in the last paragraph. It is important to note that the quantum states ψ_P are all the same since pairs of electrons are bosons and are not subject to the exclusion principle. Thus each pair has the same energy.

According to the BCS theory, v_k^2 is given by

$$v_k^2 = \frac{1}{2}\left[1 - \frac{\varepsilon_k}{\sqrt{\varepsilon_k^2 + \Delta^2}}\right] \tag{4.15}$$

where the kinetic energies of the electrons are now measured with respect to the Fermi level, i.e. $\varepsilon_k = E_k - E_F$ and where Δ is an parameter, which is approximately equal to the binding energy of the single Cooper pair, but is given by

$$\Delta = 2\hbar\omega_D \exp(-1/D(0)V) \tag{4.16}$$

The function v_k^2 is plotted in Fig. 4.13 alongside the Fermi function for a normal metal to show how the sharp discontinuity of the latter has disappeared. The condensation energy (i.e. the reduction in the total energy achieved by pairing all the conduction electrons) is given by

$$W(0) = \frac{-2D(0)(\hbar\omega_D)^2}{[\exp(2/D(0)V) - 1]} \simeq -\tfrac{1}{2}D(0)\Delta^2 \tag{4.17}$$

It will be noted that BCS makes use of the isotropic interaction energy V employed by Cooper. At $T > 0$ not all the electrons are paired but the single-particle excitations or quasi-particles are not quite the same as those in a normal metal. In the latter an electron with $|k_1| < k_F$ may be transferred into an unoccupied state with $|k_2| > k_F$ (Fig. 4.14a). An energy $(\varepsilon(k_1) + \varepsilon(k_2))$ is

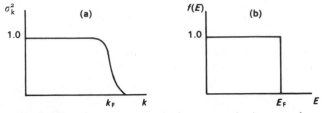

Figure 4.13 (a) Probability of pair occupancy in the superconducting ground state. (b) Fermi–Dirac distribution in the normal state at $T = 0\,\mathrm{K}$.

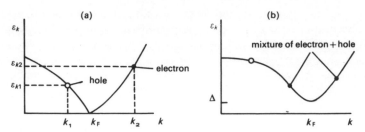

Figure 4.14 Excitation spectrum for quasiparticles in (a) a normal metal and (b) a superconductor.

required to create this electron–hole pair. Let us now see what happens when the Cooper pair $(\mathbf{k}, -\mathbf{k})$ is broken to form singly-occupied states $\mathbf{k}'$ and $-\mathbf{k}$. For this process to occur the pair state $(\mathbf{k}', -\mathbf{k}')$ must not be occupied; a probability that is given by $(1 - v_k^2)$. Because of the rounding of v_k^2 in the region near k_F, the excitations in this region are part electron and part hole. Destruction of a Cooper pair creates two quasi-particles which both have this combined electron-hole-like character. For $k \gg k_F$ where $v_k^2 \simeq 0$, the excitation is essentially an electron, while for $k \ll k_F$ where $(1 - v_k^2) \simeq 0$, the excitation is like a normal hole. The energy of a quasi-particle with wave vector $\mathbf{k}$ is found from the reduction in $W(0)$ when the pair $(\mathbf{k}, -\mathbf{k})$ is destroyed and is given by

$$\epsilon_k = (\varepsilon_k^2 + \Delta^2)^{1/2} \tag{4.18}$$

The important feature of the quasi-particle energy spectrum shown in Fig. 4.14(b) is the energy gap between the superconducting ground state and the lowest state of the excitation, although for $\varepsilon_k \gg \Delta$, $\epsilon_k \simeq \varepsilon_k$ and the quasiparticle energy is just the same as that required to excite an ordinary electron above the Fermi surface.

The last paragraphs were concerned with the behaviour of a superconductor at a temperature very close to zero when the first Cooper pair was broken, either by thermal excitation or by absorption of energy from an external source of photons or phonons. As the temperature is raised, more and more pairs are broken by thermal agitation in accordance with the laws of statistical mechanics. The superconductor can then be described qualitatively by the two-fluid model, with the quasiparticles having the properties of the normal component. The energy gap cannot remain constant, however, since occupation of states by single particles makes them unavailable for the pairing process. This leads to a reduction in both the pairing energy and the energy gap. At the transition to the normal state, the thermal energy $k_B T_c$ is approximately equal to the zero-temperature pair binding energy $W(0)$. The Gibbs' free energy in the BCS state and the normal state are found to be equal when

$$\Delta(0) = 1.76 k_B T_c = 2\hbar\omega_D \exp(-1/D(0)V) \tag{4.19}$$

Since ω_D is proportional to $M^{-1/2}$, the isotope effect follows directly from (4.19). Although the energy gap vanishes at T_c, there is little variation in $\Delta(T)$ up to $T_c/2$ and in the region near T_c, the energy gap follows the approximate relationship

$$\Delta(T) \simeq 3.1\, k_B T_c (1 - T/T_c)^{1/2} \qquad (4.20)$$

Values of the zero-temperature gap may be determined from heat capacity, electromagnetic absorption or tunnelling measurements, and for many superconductors, the ratio $\Delta(0)/k_B T_c$ lies within the range 1.6 to 2.3.

4.5 Consequences of the BCS theory

Armed with a knowledge of the structure of the ground states and of the single-particle excitation, we now have a model which can bring together almost all of the properties of the superconducting state. Many of them we have anticipated already: the link between the temperature dependence of heat capacity and the frequency dependence of surface impedance with a temperature dependent energy gap; the absence of latent heat and $\Delta(T)$ going to zero at T_c; a condensation energy $W(0)$ which may be expressed in terms of the critical field. This last example leads to a law of corresponding states for superconductors. From equations 4.4 and 4.17

$$W(0) = (G_n - G_s) = \frac{\mu_0 H_c^2}{2} = \tfrac{1}{2} D(0)\Delta^2(0)$$

Using equation 4.19 and the expression for the normal metal density of states we find

$$\frac{H_c^2}{\gamma T_c^2} = \frac{0.47}{\mu_0} \qquad (4.21)$$

where γ is the Sommerfeld constant (§3.2). Experimental values of this ratio show some scatter about the BCS value and although a more realistic choice for the electron–electron interaction would clearly modify (4.21) to account for this variation, this simple law reinforces the remarkable feature of the BCS theory that, without including any adjustable parameters, it correctly accounts for the superconducting properties of metals with widely different normal state properties.

The detailed calculations of the electrodynamic properties are quite intricate and will not be reproduced here, but within the framework of the BCS theory, it may be shown that a superconductor will expel magnetic fields and that there is a temperature-dependent penetration depth. The Pippard and London expressions for the current density are reproduced and the BCS coherence length ξ_0 is the same as that suggested by uncertainty arguments but with a value of $a = 0.18$. (It will be recalled that experiment suggested a value of $a = 0.15$).

The most striking feature of the superconducting state is, of course, the total disappearance of electrical resistance at T_c. To understand this fundamental property we first consider why a normal metal has resistance even at $T = 0$. When a current flows under the influence of an electric field, the Fermi sphere is displaced in reciprocal space as the conduction electrons gain additional momentum. An equilibrium state is maintained, while the electric field is present because the acceleration of the electrons is balanced by the deceleration produced by impurity scattering of the Bloch waves, even in the absence of thermal phonons (§3.4). In the superconducting state the electrons are paired, but of greater importance is the strongly correlated nature of the ground state. When a supercurrent flows, the total momentum of every pair is increased by an amount (say) $\hbar\delta\mathbf{k}$, i.e. the electrons which make up a pair have momenta

$$\hbar(\mathbf{k} + \delta\mathbf{k}/2, -\mathbf{k} + \delta\mathbf{k}/2)$$

To maintain the attractive interaction, the electron pair simply scatters into new states

$$\hbar(\mathbf{k}' + \delta\mathbf{k}/2, -\mathbf{k}' + \delta\mathbf{k}/2)$$

so that the overall momentum is conserved. However the total energy of the electrons, relative to the marginal ground state energy, will have been increased by

$$\delta E = \frac{n_e \hbar^2}{2m_c}\left(\frac{\delta k}{2}\right)^2$$

Once the current is established it cannot be destroyed by normal-state processes in which a single electron changes its momentum by scattering from one side of the Fermi surface to the other, by the horizontal processes described in §3.4. In the superconductor, the creation of the necessary quasiparticle by pair breaking requires a minimum energy of 2Δ. Unless the current is large, any elastic scattering process which involves a single particle excitation will further increase the energy of the electron system. As the current density is increased the situation will be reached eventually when the additional kinetic energy of the current-carrying state can be reduced by the creation of two quasiparticles. At this critical current, scattering accompanied by change of momentum becomes possible and dissipation occurs. Its value is given by

$$J_c = en_e\Delta/\hbar k_F \tag{4.22}$$

and for tin J_c is $2.10^{11}\,\text{Am}^{-2}$.

At temperatures above zero, the thermally excited quasiparticles behave like normal electrons insofar as they can be scattered and they would cause resistance if they carried a current. The remaining pairs, however, retain their zero temperature properties and cannot be scattered. Thus there can be no electric field in the superconductor to accelerate the normal electrons and the zero resistance state is maintained up to T_c, even though the critical current will fall with increasing temperature.

The resistanceless state of a superconductor reinforces the importance of the highly correlated nature of the ground-state wavefunction which is resistant to modification by single-particle processes. The coherence of the phase extends through the superconductor, a distance which may be considerable in a solenoid or a transmission line. As with helium, these properties of the wave function make possible the observation of superfluid flow on a macroscopic scale.

A more detailed explanation of the results of the tunnelling in a SIS junction (§4.3) is now possible. We noted that for both normal and superconducting junctions, the $I-V$ characteristics are influenced by the density of states in the metals on either side of the barrier. In a superconductor the energy gap in the excitation spectrum causes a displacement of states for a range Δ above and below the Fermi surface. The BCS theory shows that the missing states are concentrated immediately above the gap and $D_s(\epsilon)$ has the form

$$D_s(\epsilon) = 0 \qquad \text{for } \epsilon < \Delta \qquad (4.23)$$

$$\frac{D_s(\epsilon)}{D(0)} = \epsilon / \sqrt{\epsilon^2 - \Delta^2} \text{ for } \epsilon > \Delta.$$

with a singularity at $\epsilon = \Delta$.

$D(0)$ is again the normal metal density of states at the Fermi level. The superconducting density of states function is incorporated in Fig. 4.15 which illustrates the tunnelling processes at $T > 0$ for two different bias voltages. The shaded region represents the occupied quasiparticle states. For $V < 2\Delta/e$, the current flow is restricted to tunnelling by thermally excited particles as shown by the horizontal transition at constant energy in the Fig. 4.15(a). For $V \geqslant 2\Delta/e$, a second tunnelling process becomes possible by the destruction of a Cooper pair on the left side and the injection of a quasiparticle into each film.

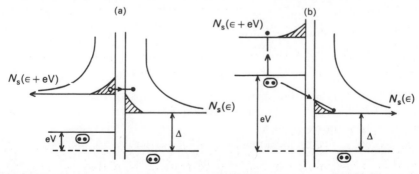

Figure 4.15 Quasiparticle tunnelling for a symmetric SIS junction at $T > 0$ K. (a) $eV < 2\Delta(T)$: a thermally-excited particle from an occupied state (shaded) tunnels across the barrier to an unoccupied state. (b) $eV \geqslant 2\Delta(T)$: a pair is destroyed on the left with the production of two quasiparticles, one of which appears on the right.

In both processes, the tunnelling current will involve products of density of states function $D_s(\epsilon)$ and $D_s(\epsilon')$ and appropriate Fermi factors. The sharp rise in I at $V = 2\Delta/e$, shown in Fig. 4.9, is a direct consequence of the singularity in $D_s(\epsilon)$ at $\epsilon = \Delta$ and this feature makes possible the direct and accurate measurement of $\Delta(T)$.

Rather different I–V characteristics are measured for non-symmetric junctions, i.e. SIS' or SIN combinations. Particularly valuable information can be obtained from derivatives of the characteristics if dI/dV (or dV/dI) is plotted against V. For the case of an SIN junction

$$dI/dV \propto D_s(\epsilon)$$

Deviations from the simple BCS expression for $D_s(\epsilon)$ are to be expected, since the true electron–electron interaction is not a constant but reflects the metal's density of phonon states. Indeed for the strong coupling superconductors like Pb and Hg in which the ratio $k_B T_c/\hbar \omega_D$ is relatively large, measurement of $D_s(\epsilon)$ can be used to determine the phonon density of states (§2.3).

4.6 Josephson effects

The question of what happens to the ground-state wavefunctions of two isolated superconductors as they are brought into a weakly coupled state was answered in a paper by Josephson in 1962, a publication that led to a Nobel prize for its author and to a sequence of remarkable effects that would eventually generate a new low-temperature technology.

If the insulating layer in a tunnel junction is made sufficiently thin, the wavefunctions overlap and the phases become locked together. Josephson's analysis tells us how much supercurrent can flow across the barrier and how the phases change when a potential difference is developed across the junction. An elegant derivation of the basic equations is given in volume III of the Feynman *Lectures on Physics* and only the relevant results will be presented here. If the coupled superconductors are linked to a battery by an external circuit, the tunnelling supercurrent that flows without a voltage appearing across the junction is given by

$$I = I_0 \sin(S_1 - S_2) = I_0 \sin S \qquad (4.24)$$

where S_1 and S_2 are the phases on opposite sides. The maximum critical current I_0, which corresponds to a phase difference of $\pi/2$, is proportional to the strength of the coupling across the barrier. I_0, which is typically $100\,\mu A$ to $1\,mA$, is determined by the dimensions of the barrier region, the materials and the temperature. When a voltage appears across the junction, the phase difference changes with time such that

$$\frac{dS}{dt} = \frac{2eV}{\hbar} \qquad (4.25)$$

It will be noted that this equation contains the basic unit of flux ($h/2e$) already encountered in the London equation. For a d.c. voltage V_0 across the junction, 4.25 can be integrated to find the phase

$$S(t) = S_0 + \frac{2e}{\hbar} V_0 t \tag{4.26}$$

and thus the current:

$$I = I_0 \sin S(t)$$

will contain a component which oscillates at a frequency v_J (the Josephson frequency), given by

$$v_j = 2eV_0/h \quad \text{or} \quad 484 \, \text{MHz/V}$$

Of the many devices in which Josephson effects are observed, three types of weak links have received most attention. As was noted earlier, a supercurrent will flow at zero voltage in an SIS tunnel junction with a barrier thickness of 10^{-9} m or less. As the film areas can be made very small, this type of Josephson junction has obvious attractions for large-scale integration applications. A point contact between two bulk superconductors, such as niobium, provides a junction whose critical current can be adjusted through the applied pressure. The point is often made from a sharpened screw with a tip radius of about $1 \, \mu$m. A thin-film microbridge is a continuous piece of superconductor containing a constriction, the dimensions of which are of the order of the coherence length.

The exact shape of the I–V characteristic depends on the type of weak link and the external circuit providing the current. In all cases, a supercurrent flows for $I \leqslant I_0$ and the phase adjusts so that no voltage appears across the junction. Typical characteristics for a thin-film tunnel junction and a point contact are shown in Fig. 4.16. It will be seen that in (a) the behaviour for $V > 0$ resembles the quasiparticle tunnelling curve of Fig. 4.9 when the junction is fed from a voltage source, but the switching follows the dotted line if a current source is used. Equation 4.26 suggests that the average d.c. current will fall to zero for

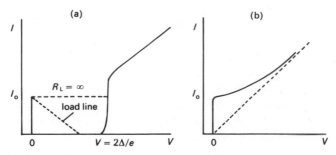

Figure 4.16 Josephson I–V characteristics for (a) a thin film tunnel junction and (b) for a point contact. In (a) the switching characteristic is determined by the load resistance R_L.

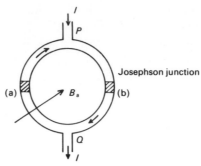

Figure 4.17   A superconducting ring containing two Josephson junctions and fed from a current source.

$V_0 > 0$. However, Josephson junctions are rather complicated non-linear devices which are shunted by capacitance and resistance and the form of their characteristics is frequently like that in Fig. 4.16(b).

Not only is the critical current of a Josephson junction reduced by increasing temperature but it is also sensitive to an applied magnetic field, being totally suppressed by field of a few mT. However a more significant field effect is observed in a superconducting loop which contains one or more Josephson junctions. In this section we will consider why the circuit shown in Fig. 4.17 behaves like a two-slit optical interferometer under the influence of a changing magnetic field. When the junctions are identical and there is no applied field, current (I) from an external source divides equally between the two possible paths so that the phase difference across the junctions is the same, i.e. $S_a = S_b$. No voltage appears between P and Q until this phase difference reaches $\pi/2$ and the critical current for the loop is simply $I_c = 2I_0$. An applied magnetic field perturbs this situation in two ways. First, a circulating current i flows along the inner surface of the loop. If there were no weak links in the circuit, this current would prevent any flux from passing through the loop and we would find that

$$Li = \Phi_a = B_a \times \text{area of loop}$$

For a single loop with mm dimensions, the self-inductance is very small and flux cancellation is not possible in the circuit containing weak links, since i will quickly exceed the critical current I_0 for the individual junctions. At this stage we will assume that $LI_0 \ll \Phi_0$ so that the flux through the loop Φ is almost exactly equal to Φ_a. The second effect of the applied field is to produce an imbalance in the phase differences across the junctions. In order to determine how the critical current I_c is modified, we must use equation 4.7a to calculate the phases S_a and S_b. Adopting the same procedure as in §4.4 for a superconducting loop, we find that

$$\Delta S = 2\pi n = \frac{2e}{\hbar} \oint \mathbf{A} \cdot d\mathbf{l} + S_a + S_b$$

or

$$S_a + S_b = 2\pi(n - \Phi_a/\Phi_0)$$

S_a and S_b are no longer equal since the currents through this junction are different, being $(i + I/2)$ and $(i - I/2)$ when measured in the same clockwise direction. This imbalance in phase can be expressed as

$$S_a = \pi(n - \Phi_a/\Phi_0) - S_0$$
$$S_b = \pi(n - \Phi_a/\Phi_0) + S_0$$

where

$$2S_0 = S_b - S_a$$

Some simple algebra allows us to eliminate i (or I) from these equations and we find that

$$I = 2I_0 \cos \pi(n - \Phi_a/\Phi_0)\sin S_0 = I_c \sin S_0$$

where the critical current for the loop I_c is now given by

$$I_c = 2I_0|\cos \pi(n - \Phi_a/\Phi_0)| = 2I_0|\cos(\Phi_a/\Phi_0)| \qquad (4.27)$$

Figure 4.18 shows how the maximum current that flows without a voltage appearing across the circuit is modulated by the flux passing through the loop. For a ring of cross-sectional area 1 cm^2, the flux periodicity of Φ_0 is equivalent to a field periodicity of 2.10^{-11} T. It is this extremely small number that makes superconducting interferometers the basic element in magnetometers and galvanometers which have the highest achievable sensitivities. The pattern of 4.18 is not observed in practice since the circulating current does provide some screening of the applied flux, so that coupled flux is

$$\Phi = \Phi_a - Li$$

The analysis of the circuit is more complicated but the resulting periodicity in the interference pattern is unchanged although the depth of modulation is reduced beacused the current minima are no longer zero. The pattern also shows a second modulation which can be viewed as a diffraction effect in a single junction. Magnetic flux passing through the area of the weak link itself cannot be neglected since it causes the critical current I_0 of the single junction

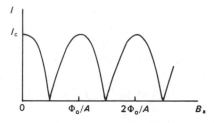

Figure 4.18 Interference pattern for a superconducting loop containing two Josephson junctions with $LI_0 \ll \Phi_0$. The area of the hole is A.

to be modulated with the periodicity Φ_0. However the field periodicity for I_0 is much larger than that for I_c because of the disparity in junction and loop areas.

We conclude this section with a brief discussion of what are called a.c. Josephson effects. When a single junction is biased at a voltage V, the Cooper pairs on opposite sides have an energy difference $2eV$. Pair tunnelling is then accompanied by the emission of energy, the frequency of the radiation being $v = 2eV/h$. Josephson devices are not useful sources of microwave radiation, however, because of the poor impedance match between the weak link (typical impedance 1 ohm) and the external circuits. Emitted power levels are typically less than 10^{-11} W. Of greater interest and practical importance is the response to a combination of a d.c. bias V_0 and a microwave electric field of frequency ω. The phase slip equation 4.25

$$\hbar \frac{dS}{dt} = 2eV_0 + 2eV_1 \cos \omega t,$$

can be integrated to find S, and I is found by substitution. Using standard mathematical identities,

$$I = I_0 \sum_{n=-\infty}^{n=\infty} (-1)^n J_n \left(\frac{2eV_1}{\hbar \omega} \right) \sin \left(\frac{2eV_0}{\hbar} - n\omega \right) t + \theta_0$$

where $J_n(x)$ are Bessel functions of the first kind and of order n and θ_0 is a constant. In general I will time-average to zero except when

$$V_0 = n\hbar\omega/2e$$

At these voltages, there is a constant voltage step in the I–V characteristic given by

$$\Delta I_n = 2I_0 J_n (2eV_1/\hbar\omega)$$

corresponding to pair tunnelling with the absorption of n photons. This result implies that the zero-voltage supercurrent will be depressed by an amount

$$\Delta I_0 = 2I_0 J_0 (2eV_1/\hbar\omega)$$

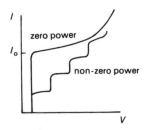

Figure 4.19   Schematic I–V characteristic for a point-contact Josephson junction irradiated with microwave power.

A schematic diagram of an $I-V$ characteristic is shown in Fig. 4.19 for increasing microwave power levels. The agreement with the Bessel functions is generally fairly good for point contacts. The importance of the a.c. Josephson effect in broad-band detection and frequency mixing will become apparent in chapter 8.

4.7 Type II superconductivity

We saw in an earlier section that the free energy of a superconductor increases in an applied field and that the Meissner state is only maintained while the free energy of the superconducting state is less than that of the normal state with total flux penetration. In some materials, penetration of magnetic flux occurs at field strengths less than the thermodynamic critical field H_{cb}. This implies that there is a lower energy state in which the material is partly superconducting with the wave function ψ (or alternatively the gap parameter) having its maximum value and with the field excluded and partly normal with zero or near-zero ψ and complete field penetration. Whether the complete or the incomplete Meissner state is stable is determined by the sign of the surface energy of the boundary which separates superconducting and normal regions. In the preceding sections it was implicitly assumed that this energy is positive and that a type I superconductor remains in the Meissner state for all field values up to the critical value H_c. A simple energy argument, however, shows that this behaviour will not be a characteristic of all superconductors.

Consider the plane boundary between the two regions in the same material. At the boundary, the magnetic flux density must rise from zero in the superconductor to H_c in the normal part over a distance equal to the penetration depth λ, as required by the condition for thermodynamic equilibrium. Likewise the superconducting wave function and hence the density of paired electrons must fall to zero over a distance equal to the coherence length ξ. Thus the formation of an $S-N$ boundary involves an increase in energy (or a loss of condensation energy):

$$\simeq \xi_0 \cdot (\mu_0 H_c^2 / 2)$$

per unit area, and a reduction in the magnetic energy:

$$\simeq \lambda \cdot (\mu_0 H_c^2 / 2)$$

per unit area. The sign of the surface energy

$$\sigma_s = (\xi - \lambda)\mu_0 H_c^2 / 2$$

will then be determined by the relative magnitudes of the coherence and penetration lengths. In many alloys and compounds (the London superconductors), $\lambda > \xi$ and the surface energy is negative. These materials exhibit type II behaviour in a magnetic field but since both λ and ξ are affected by the electron mean free path, type I elements can be made type II by appropriate alloying to reduce l.

In reality the mixed state of a type II superconductor is not made up of planar sections of normal and superconducting material. The magnetic flux penetrates the bulk in the form of cylindrical tubes or cores lying parallel to the applied field. This flux is generated by a vortex of persistent current which circulates around the core in the opposite sense to the screening currents which flow in the surface to maintain the diamagnetism of the bulk. If the superconducting wave function ψ is to fall to zero at the centre of the flux line, the effective radial dimensions of the normal core must be ξ although there is no discontinuity in either ψ or the local flux and current densities which must decay over a radial distance λ. For the formation of an isolated flux line to be energetically favourable, a condition like

$$\lambda > \xi$$

will again be necessary. A more detailed treatment modifies this expression slightly and the condition for type II superconductivity is $\kappa_G > 1/\sqrt{2}$, where κ_G is the temperature-independent Ginzburg–Landau parameter (λ/ξ). The κ_G-value increases with normal state resistance ρ and a useful approximation, in the so-called dirty limit of $l \ll \xi_0$, is due to Goodman (1962)

$$\kappa_G = \kappa_0 + 2.4 \cdot 10^6 \, \rho\gamma^{1/2}$$

where κ_0 is the pure material value and γ is again the Sommerfeld constant. In Nb and V, the coherence length is short enough, even in the absence of impurities, for their κ_G values to be greater than 0.78. These metals are therefore intrinsic type II superconductors.

The lower critical field H_{c1} is defined as the value of the applied field at which it is energetically favourable to form an isolated flux line. The argument that we have used before suggests that flux penetration will occur when

$$H_{c1} = H_{cb}/\kappa_G\sqrt{2}$$

Close to H_{c1} the isolated flux lines are separated by a distance of order λ and as the number increases, there will be a corresponding reduction in the sample magnetization below its Meissner state value. Further increases in the applied field cause the vortex lines to become more densely packed and, because the magnetic interaction between flux lines increases in importance, the stable configuration is a regular array or lattice. The upper critical field H_{c2}, when the magnetization falls to zero is reached when the lattice spacing is about ξ and the normal regions overlap. This gives a relation for H_{c2}

$$\mu_0 H_{c2} \simeq \Phi_0/\pi\xi^2$$

The arguments of the last paragraphs provide a qualitative understanding of the magnetization curve for a type II material as shown in Fig. 4.5. A very successful description of the mixed state was developed on the basis of the work of Ginzburg and Landau (1950), which in turn was based on Landau's work on second-order phase transitions. In the latter the free energy of the system is expanded as a power series of an order parameter, which is a physical property of the ordered state. Thus the theory applies in the region near the critical temperature where the order parameter is small. Ginzburg and Landau identified the macroscopic wave function as the order parameter in a superconductor. The free energy is written as a power series in ψ^2, plus terms which take account of the kinetic energy of the pairs when ψ has a spatial variation and of the magnetic field energy. The conditions for an equilibrium ordered state are found by differentiation of the free energy with respect to both ψ and $\mathbf{A}$ (the vector potential). This leads to the Ginzburg–Landau equations: the first is an equation of motion for ψ, which can be thought of as a non-linear Schrödinger wave equation

$$\tfrac{1}{2}m_o(-i\hbar\nabla - 2e\mathbf{A})^2\psi + \alpha\psi + \beta|\psi|^2\psi = 0 \qquad (4.28)$$

where α and β are the coefficients in the original expansion and the second is the same as the generalized London equation (4.7a) for the supercurrent density. Equation 4.28 can be used to make quantitative predictions about the mixed state. Since ψ must be small near H_{c2}, the non-linear term may be dropped and a value of $\mathbf{A}$ appropriate to a uniform applied field used. The result of solving the linear equation is:

$$H_{c2} = \kappa_G\sqrt{2}H_{cb}$$

As values of κ_G can be very large, we see why H_{c2} is very much greater than H_{c1}. In 1957, Abrikosov demonstrated how the non-linear term $\beta|\psi|^2\psi$ in 4.28 could be handled by perturbation methods. One result of his work was to predict that the magnetization near H_{c2} should follow

$$-M = \frac{\mu_0(H_{c2} - H_a)}{(2k_G^2 - 1)\beta_A}$$

In this expression, the parameter β_A relates to the structure of the flux line lattice. Abrikosov originally proposed a square array of vortex lines but later work by Kleiner et al. (1964) was to show that a triangular lattice as shown in Fig. 4.20a would give a more stable configuration, a result that was later confirmed by the electron micrographs of Essmann and Traüble (Fig. 4.20b).

4.8 Recent developments

Most of this chapter has been concerned with the well established properties of superconductors and in chapter 8 we will see how a new low-temperature

(a)

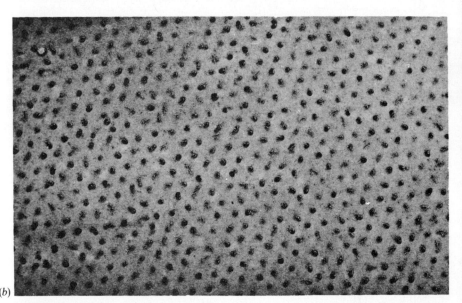

(b)

Figure 4.20 (a) Contours of constant values of $|\psi|^2$ for a triangular array of vortex lines. (After a calculation by Kleiner *et al.*, 1964.) (b) Electron micrograph of the distribution of cobalt particles on the surface of a niobium disc in the mixed state (After Essmann and Träuble, 1967; reproduced by courtesy of the authors.)

technology has been built up around them. Fundamental research in superconductivity continues but in rather esoteric fields and only a limited number of topics can be described here.

The search for new materials concentrates on increasing the superconducting transition temperature, since the upper critical field and the critical current density (§8.2) both scale to some extent with T_c, and refrigeration costs could be reduced if higher operating temperatures were possible. Two different paths could be followed. The first would aim to refine, develop or alter the structure and composition of existing materials. An example of this approach is the development of ternary compounds of Nb, Ge and Al from a knowledge of the well-known A15 compounds Nb_3Sn and V_3Ga (§8.2). The alternative route, which we will explore here, is to look for entirely new materials, for which the BCS mechanism for electron pairing may be inappropriate.

In 1964 Little proposed that it should be possible to fabricate a room-temperature superconductor from material containing chains of organic molecules. His hypothetical structure would have a spine of unsaturated carbon atoms, acting as a channel for electron conduction, and highly polarizable side groups. The oscillation of charges on the side groups replaces the lattice oscillation of the BCS superconductor by setting up regions of positive charge density through which a pair of electrons would couple. No organic superconductors with the Little structure have been found but the complex organic system tetrathiafulvalene-tetracyanoquinodimethane (TTF-TCNQ) has shown imminent superconductivity with a conductivity maximum at 58 K. Unfortunately, one-dimensional conductors which are made up of chains or stacks of large flat molecules can undergo a lattice distortion, associated with a charge density wave, and this mechanism causes TTF-TCNQ to become a semiconductor at 53 K. It was thought that high pressure might inhibit the onset of a charge density distortion and although its effect on TTF-TCNQ does actually stabilize an insulating phase, compounds of this general type, in which sulphur is replaced by selenium (TTF $\rightarrow$ TSF) and in which dimethyl or tetramethyl groups are added (TSF $\rightarrow$ TMTSF; TCNQ $\rightarrow$ DMTCNQ), remain moderately good conductors at ^{4}He temperatures when cooled under pressures of 10^4 atmospheres. For example TMTSF-DMTCNQ has a conductivity of 10^5 ohm cm at 4 K (about a factor 10^3 less than a typical copper sample) but it does not become a superconductor.

Superconductivity has been found in a number of quasi one-dimensional conducting salts based on TMTSF, with the general formula $(TMTSF)_2X$ where X is an organic ion such as NO_3, PF_6 or ClO_4, but only at temperatures of about 1 K, and in most cases a very high pressure is needed to prevent a transition to an insulating phase. However there is evidence from electron tunnelling experiments that short-range superconducting order develops at much higher temperatures, possibly as high as 40 K. The onset of this order probably involves individual chains of molecules, while the lower transition temperature, below which both zero resistance and flux expulsion are observed, determines the onset of coherence between the phase of the order parameters of adjacent chains. Thus the low-temperature/pressure phase diagram has three components, namely a one-dimensional superconductor, a three-dimensional superconductor, and an insulating phase which is possibly magnetically ordered. What mechanism can cause electron pairing in these materials is a question that has yet to be answered in full. Current ideas on this matter are well set out in the review articles of Jerome (1981 and 1982), but it would seem that neither a BCS nor a Little-type interaction is appropriate.

A rather different interpretation of enhanced superconductivity relates to the behaviour of materials when their properties are strongly perturbed by an external influence, such as microwave or ultrasonic radiation, or through the injection of a large quasi-particle current in a tunnel junction. Under certain conditions, the superconductor is driven out of thermal equilibrium into a non-equilibrium state, which may be characterized by an increase in the energy gap Δ or in the critical current or in T_c. Under a continuous perturbation, the system will reach a state of dynamic equilibrium and whether the enhancement effects are observed depends on the relaxation processes, which tend to restore the unperturbed state. The theory of gap enhancement, described by Eliashberg (1970), considers what happens to the thermally excited quasi-particles when a superconductor is irradiated by photons or phonons, with energies $h\nu < 2\Delta(T)$. Pair breaking does not occur and energy absorption takes place by inelastic scattering of quasiparticles into higher energy states, away from the gap edge. It will be recalled that $\Delta(T)$ falls with increasing temperature because states with $k \simeq k_F$, which at $T = 0$ would participate in the pairing mechanism, are singly occupied at $T > 0$. By emptying these states artificially, they are once more available to take part in the BCS interaction and the energy gap is restored towards its zero-temperature value. A comprehensive range of experiments lends support to the Eliashberg theory. Microwave photons and phonons will enhance the critical current of a Josephson junction (Wyatt et al., 1966; Tredwell and Jacobsen, 1976), results which can be interpreted as indirect evidence for gap enhancement. By irradiating an aluminium tunnel junction with 10 GHz microwave radiation, Kommers and Clarke (1977) obtained direct evidence of a power-dependent shift of the structure in the differential characteristic $(dV/dI v \cdot V)$ at the voltage corresponding to the energy gap. Resistance measurements on an Al strip (Mooij et al., 1980) confirm that the transition in a 4 GHz microwave field, although hysteretic, occurs at a temperature of > 0.02 K above the equilibrium T_c. Although these and other non-equilibrium properties may have little practical significance, they are a source of continuing interest for both theorists and experimentalists and are the subjects of a comprehensive review edited by Gray (1981).

Another situation in which changes of phase at low temperature involve different ordering processes occurs when there are both long-range magnetic ordering and superconductivity. At first sight, the coexistence of these ordered states would seem improbable, since the interaction between localized magnetic moments and conduction electrons is likely to cause depairing. Indeed, the addition of a few percent of Mn or Fe to Zn or Cd is sufficient to cause depression of T_c to zero. There are a series of ternary compounds of rare earth metals, however, such as $ReMo_6S_8$ and $ReMo_6Se_8$, in which the exchange interaction between the magnetic moments of the rare earth ions (Re) and the conduction electrons (mostly $4d$ electrons of Mo) is sufficiently weak for superconductivity and antiferromagnetism to coexist. For example $Er_{1.2}Mo_6Se_8$ becomes superconducting at $T = 6$ K, but at $T = 1.07$ K there is a λ-like anomaly in the heat capacity which is suggestive of the onset of magnetic ordering. As the material is an extreme type II superconductor, susceptibility measurements can be made in quite large magnetic fields and these indicate that the second phase transition is to the antiferromagnetic state, a result that is confirmed by neutron diffraction experiments.

When the ordered state is ferromagnetic, the lower-temperature transition is accompanied by the disappearance of superconductivity. Re-entrant behaviour is observed in $ErRh_4B_4$, which has a superconducting transition at $T = 8.7$ K, but on further cooling to $T = 0.9$ K, there is again a strong anomaly in the heat capacity as ferromagnetic order sets in. In between these temperatures the material is in an *intermediate phase* which is thought to involve the coexistence of normal ferromagnetic domains and superconducting domains with sinusoidally ordered magnetic spins. Reviews of the properties of these novel materials have been given by Maple (1978) and Tachiki (1981).

Bibliography

References

Abrikosov, A.A. *Zh. Eksp. Teor. Fiz.* **32**, 1442 (1957); translated in *Sov. Phys. J.E.T.P.* **5**, 1174 (1957).
Bardeen, J., Cooper, L.N. and Schrieffer, J.R. *Phys. Rev.* **108**, 1175 (1957).
Biondi, M.A., Garfunkel, M.P. and McCoubrey, A.O. *Phys. Rev.* **108**, 495 (1957).
Biondi, M.A. and Garfunkel, M.P. *Phys. Rev. Lett.* **2**, 143 (1959).
Cooper, L.N. *Phys. Rev.* **104**, 1189 (1956).
Deaver, B.S. and Fairbank W.M. *Phys. Rev. Lett.* **7**, 43 (1961).
Eliashberg, G.M. *Zh. Eksp. Teor. Fiz. Pisma.* **11**, 186 (1970); translated in *Sov. Phys. J.E.T.P. Lett.* **11**, 114 (1970).
Essmann, U. and Träuble, H. *Phys. Lett.* **24A**, 526 (1967).
Feynman, R.P., Leighton, R.B. and Sands, M. *Lectures on Physics Vol. III*. Addison–Wesley, Reading (Mass.) (1965).
Fröhlich, H. *Phys. Rev.* **79**, 845 (1950).
Giaever, I. *Phys. Rev. Lett.* **5**, 147 and 464 (1960).
Ginzburg, V.L. and Landau, L.D. *Zh. Eksp. Teor. Fiz.* **20**, 1064 (1950).
Goodman, B.B. *IBMJ Res. Development* **6**, 63 (1962).
Jerome, D. *Physica* **109** (B + C), 1447 (1981): also *Scientific American* **247**, 50 (1982).
Josephson, B.D. *Phys. Lett.* **1**, 251 (1962).
Kleiner, W.H., Roth, L.M. and Autler, S.H. *Phys. Rev.* **A133**, 1226 (1964).

Kommers, T.M. and Clarke, J. *Phys. Rev. Lett.* **38**, 1091 (1977).
Little, W.A. *Phys. Rev.* **A134**, 1416 (1964).
London, F. and H. *Proc. Roy. Soc.* **A149**, 72 (1935) and *Physica* **2**, 341 (1935).
Maple, M.B. *J. de Physique* **C6**, 1375 (1978).
Meissner, W. and Oschenfeld, R. *Naturwissenschaften* **21**, 787 (1933).
Mooij, J.E., Lambert, N. and Klapwijk, T.M. *Sol. State. Comm.* **36**, 585 (1980).
Phillips, N.E. *Phys. Rev.* **134**, 385 (1964).
Pippard, A.B. *Proc. Roy. Soc.* **A216**, 547 (1953); also Faber, T.E. and Pippard, A.B. *op. cit.*, **A231**, 336 (1955).
Tachiki, M. *Physica* **109** (**B & C**), 1699 (1981).
Tredwell, T.J. and Jacobsen, E.H. *Phys. Rev.* **B13**, 2931 (1976).
Van Harligen, D.J. *Physica* **109** (**B & C**), 1710 (1981).
Wyatt, A.F.G., Dmitriev, V.M., Moore, W.S. and Sheard, F.W. *Phys. Rev. Lett.* **16**, 1166 (1966).

Further reading

Bleaney, B.I. and Bleaney, B. *Electricity and Magnetism.* Oxford University Press, Oxford (1976).
Gray, K.E. *Non-equilibrium Superconductivity, Phonons and Kapitza Boundaries.* Plenum Press, New York (1981).
London, F. *Superfluids*, Vol. I, (*Superconductivity*) Wiley, New York (1950); reprinted by Dover, New York (1961).
London, F. *Superfluids.* Vol. II, (*Superfluid Helium*) Wiley, New York (1954); reprinted by Dover, New York (1964).
Rose-Innes, A.C. and Rhoderick, E.H. *Introduction to Superconductivity.* Pergamon, London (1977).
Tilley, D.R. and Tilley, J. *Superfluidity and Superconductivity.* Van Nostrand Reinhold, London (1974) and new edition (Adam Hilger, forthcoming).
Tinkham, M. *Introduction to Superconductivity.* McGraw-Hill, New York (1975).

5   Liquid helium-4

5.1 Influence of Bose–Einstein statistics

It is immediately evident from the measured properties of liquid ^{4}He that something very remarkable indeed happens at the lambda transition. All the thermodynamic properties of the liquid (or their derivatives) undergo discontinuous changes at the transition temperature T_λ ($= 2.17$ K at the saturated vapour pressure). The heat capacity is plotted, as one example, in Fig. 5.1 (and see also the velocities of sound in Fig. 5.12). For temperatures below T_λ, liquid ^{4}He behaves in many ways as though it was actually a mixture of *two* separate fluids, one of which is devoid of entropy and viscosity (see §5.2). No such behaviour occurs above T_λ, where the properties of the liquid are quite ordinary by comparison, being rather similar to those expected of a dense classical gas of hard spheres.

The explanation for this strikingly peculiar behaviour was given by London (1938). He proposed that, in view of its extremely low density and viscosity (see §1.5), liquid helium might appropriately be treated as a gas, albeit a rather non-ideal one. Being made up from an even number of fundamental particles (2 protons, 2 neutrons, 2 electrons, all of which have spin $\frac{1}{2}\hbar$), the ^{4}He atom has no resultant intrinsic angular momentum. It is therefore a boson, with a symmetric overall wavefunction. There is thus no restriction on the number of ^{4}He atoms allowed to occupy any particular quantum state. If the interatomic forces can to a first approximation be ignored, the assembly will be described by Bose–Einstein statistics, so the probability f_i that a state of energy E_i will be occupied is specified (see, for example: Mandl, 1971) by the Bose–Einstein distribution function

$$f_i = \frac{1}{e^{(E_i - \mu)/k_B T} - 1} \tag{5.1}$$

where μ is the chemical potential. London suggested that the lambda transition in liquid ^{4}He might be the analogue of *Bose–Einstein condensation*, a curious phenomenon predicted to occur in an ideal gas of bosons when cooled to a low enough temperature. The expectation is that, as the

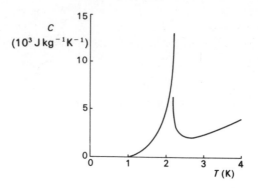

Figure 5.1   The heat capacity C of liquid ^{4}He as a function of temperature T (after Atkins, 1959).

temperature is gradually reduced, the ground state will quite suddenly become occupied by a significant fraction of all the particles in the assembly.

A simple argument suffices to demonstrate why this must be so. The permitted translational energy levels E_i for a set of non-interacting particles of mass m in a box of volume V (subject to cyclic boundary conditions) are

$$E_i = \frac{h^2}{2mV^{2/3}}(n_1^2 + n_2^2 + n_3^2) \tag{5.2}$$

where n_1, n_2, n_3 are any positive or negative integers. The number of particles occupying any particular energy level

$$n_i = f_i g_i = \frac{g_i}{e^{(E_i - \mu)/k_B T} - 1} \tag{5.3}$$

where g_i is the degeneracy (the number of permitted states) corresponding to the energy E_i. The total number of particles in the assembly may therefore be written

$$N = \sum_i \frac{g_i}{e^{(E_i - \mu)/k_B T} - 1}, \tag{5.4}$$

thereby defining the value of μ. The levels are, of course, extremely closely spaced in energy, and so it is convenient to replace the summation in (5.4) with an integration. In doing so, g_i must be replaced by the appropriate density of states function

$$D(E) = \frac{2\pi}{h^3} V (2m)^{3/2} E^{1/2}. \tag{5.5}$$

It turns out to be of crucial importance to note that, although (5.5) will be an excellent approximation when (as is usually the case) the energy difference between successive levels is negligible compared to their absolute magnitude, it is bound to be less satisfactory at very low energies. In particular, the form of $D(E)$ in (5.5) takes *no account at all* of the existence of the ground state, given by $n_1 = n_2 = n_3 = 0$ in (5.2). Thus, in changing to an integral formulation of (5.4), we must write

$$N_e = \int_0^\infty \frac{D(E)\,dE}{e^{(E-\mu)/k_B T} - 1}$$

or

$$N_e = \frac{2\pi V}{h^3}(2m)^{3/2} \int_0^\infty \frac{E^{1/2}\,dE}{e^{(E-\mu)/k_B T} - 1} \tag{5.6}$$

where N_e refers to the number of particles in excited states. Next, we note that μ cannot be positive: if it were greater than zero then, according to (5.1), there would exist a range of E_i values for which the probability of occupation was negative; and this possibility we reject as being physically meaningless. Therefore, (5.6) implies that

$$N_e \leqslant \frac{2\pi V}{h^3}(2m)^{3/2} \int_0^\infty \frac{E^{1/2}\,dE}{e^{E/k_B T} - 1}$$

or, with a change of variable to $x = E/k_B T$,

$$N_e \leqslant \frac{2\pi V}{h^3}(2mk_B T)^{3/2} \int_0^\infty \frac{x^{1/2}\,dx}{e^x - 1}.$$

The integral is now in a standard form taking the value 2.315, so that

$$N_e \leqslant \frac{41.1\,V}{h^3}(mk_B T)^{3/2}. \tag{5.7}$$

This is a very surprising result. It clearly implies that the total number of particles in excited states must be less than a quantity which itself tends to zero

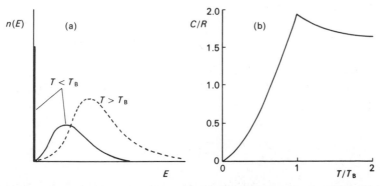

Figure 5.2 Some properties expected of an ideal Bose–Einstein gas near its condensation temperature T_B: (a) the energy distribution function $n(E)$; (b) the specific heat C in units of the gas constant R, plotted against reduced temperature.

F

as $T^{3/2}$. As the temperature is reduced, a crisis is bound to occur at some critical temperature T_B when the right-hand side of (5.7) has fallen to a value smaller than the total number of particles in the assembly. The inescapable conclusion is that, for $T < T_B$, the rest of the particles must be 'forced' into the zero-momentum ground state since there is nowhere else for them to go. The distribution function $n(E)$ must therefore take the exceedingly odd form sketched in Fig. 5.2(a). For $T > T_B$, $n(E)$ approximates to the usual Maxwell–Boltzmann shape found at room temperature. Below T_B, however, an additional spike appears at $E = 0$, corresponding to a macroscopic number of particles condensing into the ground state. It is these latter particles which constitute the Bose–Einstein condensate referred to above in §1.6.

For $T \leqslant T_B$ it can be shown (London, 1938) that $\mu \simeq 0$, implying that N_e stays as close as possible to its maximum allowed value. To a good approximation, therefore, we can replace (5.7) with

$$N_e = \frac{41.1\,V}{h^3} (mk_B T)^{3/2} \tag{5.8}$$

and can define T_B by setting

$$N = \frac{41.1\,V}{h^3} (mk_B T_B)^{3/2}. \tag{5.9}$$

From (5.8) and (5.9) we obtain

$$\frac{N_e}{N} = \left(\frac{T}{T_B} \right)^{3/2} \tag{5.10}$$

so that

$$\frac{N_0}{N} = 1 - \left(\frac{T}{T_B} \right)^{3/2}. \tag{5.11}$$

Also, from (5.9),

$$T_B = \frac{h^2}{11.9 m k_B} \left(\frac{N}{V} \right)^{2/3}. \tag{5.12}$$

Here, N/V is, of course, the number of particles per unit volume; and N_0/N is the proportion in the ground state. If we make the usual assumption that the particles in the excited states have an average energy of $c. \frac{3}{2} k_B T$, (5.10) implies that the molar internal energy U for $T \leqslant T_B$ will be given approximately by

$$U \sim \tfrac{3}{2} RT(T/T_B)^{3/2} \tag{5.13}$$

and the molar heat capacity for $T \leqslant T_B$ by

$$C = \left(\frac{\partial U}{\partial T} \right) \sim \tfrac{15}{4} R(T/T_B)^{3/2} \tag{5.14}$$

A fuller treatment (London, 1938) gives

$$C = 1.93R(T/T_B)^{3/2}. \tag{5.15}$$

The calculated specific heat is plotted as a function of temperature in Fig. 5.2(b). We note in passing that C tends to zero at absolute zero, in accordance with the requirements of the Third Law of Thermodynamics (see §1.2).

What relevance do these various considerations have for liquid ^{4}He? In view of its interatomic forces which are, by definition, entirely absent for the ideal Bose–Einstein gas, one must clearly proceed with caution; but, since the forces are weak and in the light of the discussions above and in §1.5, it is not at all unreasonable to hope that there will be some correspondence between the ideal-gas model and the experimentally observed properties of the liquid. Substitution of the atomic mass and number density for liquid ^{4}He ($N/V = 2.2 \times 10^{28}\,\text{m}^{-3}$ under the saturated vapour pressure at low temperatures) into (5.12) shows that a Bose–Einstein condensation in an ideal gas of the same density would occur at $T_B = 3.1$ K. This is encouragingly close to the temperature (2.17 K) at which the lambda transition is observed in liquid ^{4}He. Although the shape of the measured specific-heat anomaly (Fig. 5.1), which actually approximates to a logarithmic infinity (Ahlers, 1976), is rather different from the finite cusp predicted for the ideal gas (Fig. 5.2b), and although the observed pressure dependence of T_λ (Fig. 1.9a) is in the opposite direction to that predicted by (5.12) for T_B, it is hard to avoid the conclusion that the lambda transition does, in fact, correspond to a Bose–Einstein

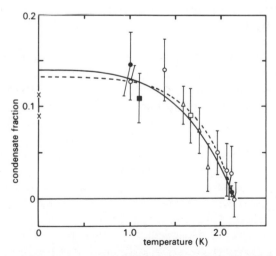

Figure 5.3   The condensate fraction of liquid ^{4}He (Sears *et al.*, 1982; Campbell, 1983). The Xs are theoretical predictions and the other points and the dashed curve represent experimental data from various sources. The full curve is a theoretical temperature-dependence fitted to the data points, enabling a $T = 0$ value of $14 \pm 2\%$ to be extrapolated.

condensation, albeit in a modified form because of the presence of the interatomic forces.

Confirmation of these ideas has recently been provided through the direct experimental detection and measurement of the condensate fraction in HeII (Sears *et al.*, 1982) based on a neutron-scattering technique. The results (Fig. 5.3) show that the condensate fraction at $T = 0$ is $c.$ 14%, rather than the 100% expected of an ideal gas. This is, however, an expected consequence of the finite interatomic forces, which cause some of the particles to occupy what are called *depletion levels*, even at $T = 0$, depleting the fraction of particles in the condensate. Further confirmation of the existence of the condensate has been provided by Campbell (1983), who succeeded in deducing the condensate fraction from surface-tension measurements (dashed curve in Fig. 5.3), an entirely different experimentally-based approach, yielding excellent agreement with the neutron-scattering data.

Although Bose–Einstein statistics have thus been shown to play a vital role in determining the low-temperature behaviour of liquid ^{4}He, we will find that, presumably because of the interatomic forces, there is no simple straightforward correspondence between the actual properties of HeII and those predicted for an ideal Bose–Einstein gas.

5.2 Two-fluid properties

Measurements of the physical properties of HeII reveal immediately that it is, to say the least, a very peculiar liquid. We will first describe some of the more striking of these properties. Then we will discuss how the phenomenological two-fluid model has been able not only to correlate the various observations within a consistent framework, but also to predict the existence of qualitatively new phenomena.

HeII is virtually a superconductor of heat, as may readily be demonstrated through observation of a dewar of liquid helium while it is being pumped to reduce its temperature past T_λ. The liquid bubbles and boils violently for $T > T_\lambda$ but, as T_λ is passed, all the bubbling and commotion suddenly ceases. This is because, on account of the huge effective thermal conductivity, the local temperature fluctuations normally responsible for bubble nucleation are dissipated too rapidly for any bubbles to be able to form: evaporation still takes place from the liquid, but only from its surface.

The viscosity η of the liquid takes different apparent values (Fig. 5.4 a, b) depending on how it is measured. It is always exceedingly small (comparable with that of air at STP) and, in many experimental situations (e.g. Fig. 5.4a) immeasurably so, whence the appellation *superfluid* in referring to HeII. The liquid tends to escape from any open-topped vessel in which it is placed, through the formation of a mobile superfluid film which creeps rapidly up and over the sidewalls. We discuss the superfluid film in more detail in §5.6.

It is found that temperature changes often occur when HeII flows through

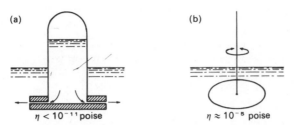

(a)

(b)

$\eta < 10^{-11}$ poise $\eta \approx 10^{-5}$ poise

Figure 5.4 Two different methods of attempting to measure the apparent viscosity η of HeII (after Atkins, 1959): (a) flow through a very narrow channel (between optically flat plates); (b) damping of an oscillating disk. As indicated, very different answers are obtained.

narrow channels. For example, if the liquid is allowed to flow out from a thermally insulated vessel through the interstices of a tightly packed powder (Fig. 5.5a), the temperature of the remaining liquid is found to rise. Using an experimental arrangement of the type shown in Fig. 5.5(b), it may be demonstrated that such temperature changes are in fact *reversible*. A very fine capillary tube is used to provide a narrow channel connecting two reservoirs of HeII. If the level in one reservoir is depressed, so that liquid flows through the capillary, it is observed that its temperature increases. Simultaneously, the temperature of the other reservoir decreases. If the process is then reversed it is found that, as the liquid flows back again, the two reservoirs return once more to their starting temperatures. Evidently, whatever liquid flows along the capillary in this experiment is not bulk HeII.

Another intriguing and highly illuminating experiment can be performed using the arrangement shown in Fig. 5.6(a). Again, an extremely fine capillary tube is used and it connects the HeII in an outer reservoir to that inside an open-topped cylinder. If the temperature of the liquid inside the cylinder is increased slightly by means of the immersed heater, liquid flows through the capillary and causes the level inside the cylinder to rise until it reaches an

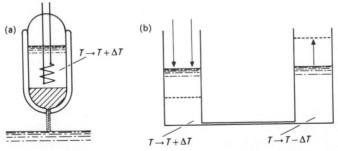

(a)   (b)

$T \rightarrow T + \Delta T$

$T \rightarrow T + \Delta T$ $T \rightarrow T - \Delta T$

Figure 5.5 Temperature changes are associated with the flow of HeII through narrow channels. (a) When HeII drains away through a plug of closely packed powder (shaded) the temperature of the remaining liquid (as measured by the immersed thermometer) rises (Daunt and Mendelssohn, 1939). (b) A similar effect occurs when HeII passes along a fine capillary tube connecting two reservoirs, and is found to be *reversible*.

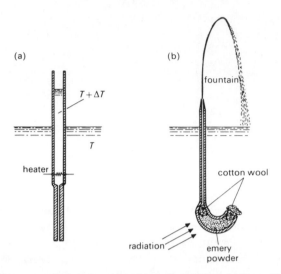

Figure 5.6   The thermomechanical (fountain) effect in HeII (Allen and Jones, 1938). (*a*) When the temperature of the HeII in the vessel is increased slightly above that of the bath, liquid flows in through the capillary tube. (*b*) The same phenomenon (using radiative heating in the case illustrated) can be used to create a dramatic fountain of liquid helium.

equilibrium position which depends on the outside and inside temperatures. A more dramatic rendering of the same phenomenon can be afforded by the so-called fountain experiment of Fig. 5.6(*b*). The narrow interstices between the grains of black carborundum power replace the capillary of Fig. 5.6(*a*); and the powder also serves as a heater by absorbing energy when a light is shone on it. Consequently, the inner liquid level rises until a jet is forced out through the narrow orifice at the top of the vessel. The resultant fountain may issue with considerable force and readily reaches a height of 300 mm or more.

It is evident from these experiments that, quite contrary to the behaviour of other liquids, a very close relationship exists in equilibrium between pressure differences and temperature differences in HeII. When a temperature difference is imposed, a corresponding pressure difference is created (Fig. 5.6), and this is known as the *thermomechanical effect*. The inverse effect, where a temperature difference results from an imposed pressure difference (Fig. 5.5*b*), is known as the *mechanocaloric effect*.

HeII possesses numerous other unusual and interesting properties, in addition to the examples mentioned above, some of which we treat below in succeeding sections of this chapter. First, however, we will introduce the *two-fluid model*, which provides a conceptual framework within which most of the strange properties of the liquid can be understood, or at least correlated with each other. The treatment will be essentially phenomenological, modelled upon that presented by Landau (1941). We will assume that we are dealing with HeII in the linear regime: that is, all velocities, temperature gradients, pressure gradients and so on will be assumed to be not too large (and the physical significance of the latter phrase will become evident in due course). The postulates upon which the model is constructed can be summarized as follows.

1. HeII behaves as though it consists of two separate fluids: a *normal fluid* component and a *superfluid* component.
2. The two fluids interpenetrate freely, passing through each other entirely without interaction.
3. The density ρ of the liquid as a whole is made up of the sum of the densities of the two components, so that

$$\rho = \rho_n + \rho_s.$$

We will be using the subscripts n and s to denote quantities characteristic, respectively, of the normal fluid and superfluid. We will find that $\rho_n = 0$ at $T = 0$, and $\rho_n = \rho$ at $T = T_\lambda$.
4. The superfluid component carries no entropy and experiences no flow resistance whatever; that is, its viscosity is identically zero and no turbulence can be created in it (but cf. §5.5). The latter property can conveniently be specified by requiring that

$$\mathbf{V} \wedge \mathbf{v}_s = 0 \tag{5.16}$$

where $\mathbf{v}_s$ is the velocity of the superfluid.
5. The normal fluid component carriers the whole entropy content S of the liquid and it possesses a finite viscosity η_n.

It will be realized that these postulates are consistent with the various experimental results discussed above and, indeed, that they were arrived at with those (and other) experimental observations specifically in mind. For example, the viscosity of HeII when measured through flow in a very narrow channel (Fig. 5.4a) will appear to be zero because only the superfluid will be able to move; whereas the oscillating disk (Fig. 5.4b) will experience drag from the surrounding normal fluid (although none, of course, from the superfluid component). Similarly, when a pressure differential is applied to the reservoirs of Fig. 5.5, only the superfluid component (carrying zero entropy) passes through the capillary, because the normal fluid is effectively 'clamped' by its own viscosity. Consequently, the entropy *density* of the depleted reservoir must increase while that of the augmented reservoir decreases, corresponding to the temperature changes actually observed.

These ideas can be developed quantitatively by deriving the relevant equations of motion for the two-fluid system. They are sometimes known as the *thermohydrodynamical equations* because of the peculiar necessity of intermingling thermodynamics and hydrodynamics in their derivation. First, we will state the equations in their simplest form. Then we will discuss the physical significance of each of them and will indicate briefly how they are derived. If viscous effects and terms which are second-order in the velocities are ignored, the set of equations may be written:

$$\rho = \rho_n + \rho_s \tag{5.17}$$

$$\mathbf{j} = \rho_n \mathbf{v}_n + \rho_s \mathbf{v}_s \tag{5.18}$$

$$\mathbf{\nabla} \cdot \mathbf{j} = -\frac{\partial \rho}{\partial t} \tag{5.19}$$

$$\mathbf{\nabla} \cdot (\rho S \mathbf{v}_n) = -\frac{\partial (\rho S)}{\partial t} \tag{5.20}$$

$$\frac{\partial \mathbf{j}}{\partial t} = -\nabla P \tag{5.21}$$

$$\frac{\partial \mathbf{v}_s}{\partial t} = S \nabla T - \frac{1}{\rho} \nabla P \tag{5.22}$$

Equations (5.17) and (5.18) represent succinctly the principal premises of the model: that the density ρ and momentum $\mathbf{j}$ of the liquid are each composed of the sum of contributions from normal and superfluid components, each of which has its own velocity field. Equation (5.19) is the usual hydrodynamic equation of continuity for the liquid as a whole and has the effect of ensuring conservation of mass. Equation (5.20) is a second equation of continuity which ensures conservation of entropy, on the assumption that viscous effects are negligible and that the motions of the two fluids are reversible. Equation (5.21) is Euler's equation (from hydrodynamics) in which terms quadratic in the velocities have been ignored and where it is again assumed that viscous effects are negligible. Equation (5.22), the acceleration equation for the superfluid component, can be derived on the basis of thermodynamic arguments. Its physical significance lies in its statement that, unlike ordinary fluids, the superfluid is as liable to be accelerated by temperature gradients as by pressure gradients (the net acceleration resulting, in fact, from the gradient in chemical potential). Detailed discussions of the derivation of this and of the other equations of the two-fluid model will be found in standard texts on liquid helium: for example, Wilks (1967).

We now consider in rather more detail how this peculiar but conceptually quite simple picture is able to account for so many of the unexpected properties of HeII. It will be noted that actual numerical values of quantities such as ρ_n and ρ_s are not predicted by the model, but must be obtained by experiment. Once they have been determined, however, the model can be used to make a variety of quantitative predictions. These can then be checked experimentally, thereby providing a stringent test of the applicability of the model itself.

The normal fluid density was determined as a function of temperature by Andronikashvili (1946) using a technique which also provided a particularly convincing demonstration of the division of the liquid into two components.

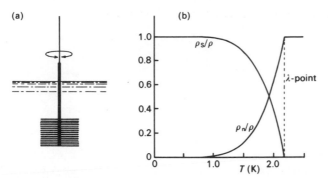

Figure 5.7 Andronikashvili's (1946) experiment, demonstrating the effective division of HeII into normal and superfluid components: (a) the immersed stack of disks, used as a torsion pendulum; (b) the deduced normal and superfluid densities, ρ_n and ρ_s, divided by the total density ρ of the liquid, as functions of temperature T.

He measured the period of an ingeniously designed torsion pendulum immersed in the liquid (Fig. 5.7a). The bob of the pendulum consisted of a stack of thin metal disks spaced about 0.2 mm apart from each other. Because of the close spacing (much less than the viscous penetration depth), normal fluid between the disks was constrained to oscillate back and forth with the bob. It therefore increased the bob's effective moment of inertia and thence, also, the period of oscillation. The (inviscid) superfluid component, by contrast, remained stationary and hence had no effect either on the moment of inertia or on the period. Thus, by measurement of the period of the pendulum as a function of temperature it was possible to deduce $\rho_n(T)$. Because the bulk density ρ of the liquid was known, values of $\rho_s(T)$ immediately followed, through use of (5.17). The results are shown in Fig. 5.7(b). It may be noted that, as expected, $\rho_s \to 0$ and $\rho_n \to \rho$ as $T \to T_\lambda$. Below c. 1 K, the liquid is composed almost exclusively of the superfluid component.

The enormous thermal conductivity of HeII may readily be understood in terms of the unusual form of convection (Fig. 5.8) made possible by the two-fluid system. Superfluid is converted to normal fluid at the heater, flows away

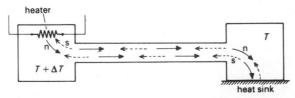

Figure 5.8 Sketch to indicate the peculiarly effective mode of thermal conduction in HeII. Heat generated at the heater converts superfluid into normal fluid, which flows (full arrows) to the heat sink. Here, it gives up all of its entropy in becoming re-converted to superfluid, which returns (dashed arrows) once more to the heater.

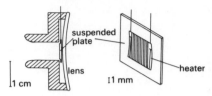

Figure 5.9 Apparatus used by Hall (1954) to study the momentum flux associated with heat flow in HeII.

from it and, in giving up its entropy to the heat sink, is converted back to superfluid again; and so on. The process is a particularly efficient one: first, because the 'hot' and 'cold' streams pass effortlessly through each other, without the usual necessity of establishing a circulating convection current driven gravitationally by small density differences; and, secondly, because the 'hot' streams gives up *all* of its entropy to the heat sink and not just a small fraction of it as in conventional convective heat transfer.

It may be noted that, because the transport of heat in HeII occurs through the counterflow of the two fluids, there will be associated fluxes of momentum. Their existence was demonstrated by Kapitza's famous jet-propelled spider (1941), and the effect has been investigated quantitatively by Hall (1954), who detected the deflection of a suspended glass plate with a wire-wound heater on one side (Fig. 5.9). The low thermal conductivity of the plate ensured that virtually all the heat moved to the left in the diagram. The flux associated with each fluid is of the form $\rho \mathbf{v} \cdot \mathbf{v}$, so that the net reaction pressure on the heat source will be

$$P_r = \rho_n v_n^2 + \rho_s v_s^2. \tag{5.23}$$

The heat flow per unit area,

$$W = \rho S T v_n.$$

Assuming that the mass flow is zero, $\rho_n v_n = -\rho_s v_s$, whence

$$P_r = \frac{\rho_n W^2}{\rho_s \rho T^2 S^2} = \frac{W^2}{u_2^2 \rho C T} \tag{5.24}$$

where u_2 is the velocity of second sound (see §5.3) and C is the specific heat. The resultant deflection of the plate, measured by the change in the Newton's rings system between it and the adjacent lens, was found to be in excellent agreement with this relation.

The thermomechanical effect (Fig. 5.6) can be thought of as a peculiar form of osmosis, with the superfluid acting as a solvent for the normal fluid solute, and the capillary or superleak playing the role of a semipermeable membrane. Thus, raising the temperature of the inner reservoir of HeII is equivalent (Fig. 5.6b) to increasing the concentration of solute (normal fluid) there, so that solvent (superfluid) enters to 'try' to dilute it. The final level difference (Fig.

5.6a) can be found from equation (5.22), with $\partial v_s/\partial t = 0$ for equilibrium, so that

$$\Delta P = \rho \int_{T_C}^{T_H} S(T)dT \qquad (5.25)$$

where T_C and T_H are the outer and inner temperatures respectively. Careful measurements of ΔP and $\Delta T = T_H - T_C$, taken in conjunction with the known value of ρ and values of $S(T)$ derived from specific heat measurements, show that (5.25) is followed very closely over a wide range. When ΔT is small, (5.25) may more conveniently be approximated by the simpler form

$$\Delta P = \rho S(T)\Delta T \qquad (5.26)$$

which is usually known as H. London's equation.

The London equation has been used as the basis of a flow technique for the measurement of η_n. As noted above, η_n cannot be measured by observing bulk flow of HeII; but the pressure gradient in thermal counterflow within a *closed* system is directly proportional to η_n which can thereby be measured. Two reservoirs of HeII are connected by a capillary tube of length l and of a radius a which is sufficiently large to permit a finite flow of normal fluid, as sketched in Fig. 5.10(a). The volume rate of flow of normal fluid, carrying heat $\dot{Q}$ from the heater to the heat sink at T_C, will be given by Poiseuille's equation

$$\dot{V}_n = \frac{\pi \Delta P a^4}{8\eta_n l} \qquad (5.27)$$

where ΔP is the pressure difference between the reservoirs. Once equilibrium has been established, the heat flow may be expressed in terms of the entropy, average temperature T and flow rate as

$$\dot{Q} = \rho S \dot{V}_n T. \qquad (5.28)$$

Eliminating $\dot{V}_n$ and ΔP (which are not easily measured) between (5.26), (5.27) and (5.28), we obtain

$$\dot{Q} = \frac{\pi T(\rho S)^2 a^4}{8\eta_n l}\Delta T. \qquad (5.29)$$

All the quantities in (5.29) except η_n are known or easily determined, so that measurement of the ΔT resulting from a given $\dot{Q}$ at different temperatures enables $\eta_n(T)$ to be found. The results obtained by Brewer and Edwards (1959) using this technique (Fig. 5.10b) are in good agreement with values derived from the damping of oscillating disks and spheres and of vibrating wires. The normal fluid viscosity rises as T is reduced below 1.8 K, but it must be borne in mind that the proportion of the liquid which is normal fluid, ρ_n/ρ, is falling extremely rapidly (Fig. 5.7b) in this temperature range.

Although the two-fluid model describes most aspects of the behaviour of

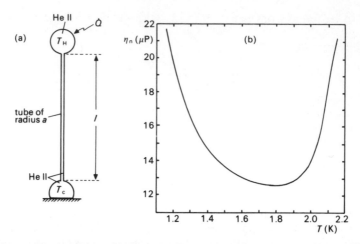

Figure 5.10 (*a*) Sketch (highly simplified) of the apparatus used by Brewer and Edwards (1959) to measure the viscosity η_n of normal fluid flowing through a capillary in response to a heat current $\dot{Q}$. (*b*) The deduced values of η_n, as a function of temperature T.

HeII remarkably well, it is important always to remember that there can be no suggestion of two physically distinct materials being present. HeII is an assembly of ^{4}He atoms, all of which are identical. As noted above, however, the liquid certainly behaves *as though* it were a mixture of two fluids. This simple idea will form the basis for the further discussions of HeII which follow.

Where velocities, or gradients in T and P, become too large, quantized vortices (see §5.5) are created in the superfluid component. These result (see §5.7) in the development of a so-called mutual friction between the two components whenever their relative velocity is non-zero, in partial violation of the second premise (above) of the two-fluid model. Such effects can readily and properly be accommodated, however, by introducing appropriate additional terms into the equations of motion: details will be found in, for example, Wilks (1967).

5.3 Wave propagation

A particularly interesting feature of HeII lies in its ability to support more than one type of sound wave. In this section we discuss the modes, known as first and second sound, which propagate in bulk HeII and, in less detail, fourth sound, which propagates through HeII held in very narrow channels. Waves on the creeping superfluid film, known as third sound, are described in §5.6.

We will seek wave-like solutions of the thermohydrodynamical equations. Eliminating **j** between (5.19) and (5.21) we can obtain

$$\frac{\partial^2 \rho}{\partial t^2} = \nabla^2 P. \tag{5.30}$$

Next we eliminate $\mathbf{v}_s$, $\mathbf{v}_n$, ρ and P between the remaining equations of motion. If we eliminate P between (5.21) and (5.22), and then take the divergence of the resultant equation,

$$\rho_n \frac{\partial}{\partial t} \nabla \cdot (\mathbf{v}_n - \mathbf{v}_s) = -\rho S \nabla^2 T. \tag{5.31}$$

Eliminating $\mathbf{j}$ between (5.18) and (5.19), and using (5.17)

$$\nabla \cdot (\mathbf{v}_n - \mathbf{v}_s) = \frac{1}{\rho_s} \left(\frac{\partial \rho}{\partial t} + \rho \nabla \cdot \mathbf{v}_n \right)$$

whence, using (5.20),

$$\nabla_0 (\mathbf{v}_n - \mathbf{v}_s) = -\frac{\rho}{\rho_s S} \frac{\partial S}{\partial t}. \tag{5.32}$$

Inserting (5.32) in (5.31) we obtain

$$\frac{\partial^2 S}{\partial t^2} = \frac{\rho_s S^2}{\rho_n} \nabla^2 T. \tag{5.33}$$

Equations (5.30) and (5.33), taken together, completely specify the propagation of (small amplitude) sound in bulk HeII. It is convenient, and also sufficient, to describe conditions in the liquid in terms of the departures from their equilibrium values of *two* of the variables: we choose ρ and S. Then any departures from the equilibrium values of P and T can be written in the forms

$$\delta P = \left(\frac{\partial P}{\partial \rho} \right)_s \delta \rho + \left(\frac{\partial P}{\partial S} \right)_\rho \delta S, \qquad \delta T = \left(\frac{\partial T}{\partial \rho} \right)_s \delta \rho + \left(\frac{\partial T}{\partial S} \right)_\rho \delta S$$

so that (5.30) and (5.33) become

$$\frac{\partial^2 \rho}{\partial t^2} = \left(\frac{\partial P}{\partial \rho} \right)_s \nabla^2 \rho + \left(\frac{\partial P}{\partial S} \right)_\rho \nabla^2 S \tag{5.34}$$

$$\frac{\partial^2 S}{\partial t^2} = \frac{\rho_s S^2}{\rho_n} \left[\left(\frac{\partial T}{\partial \rho} \right)_s \nabla^2 \rho + \left(\frac{\partial T}{\partial S} \right)_\rho \nabla^2 S \right]. \tag{5.35}$$

We now look for plane-wave solutions of (5.34) and (5.35) of the standard form

$$\rho = \rho_0 + \rho' \exp[i\omega(t - x/u)], \quad S = S_0 + S' \exp[i\omega(t - x/u)] \tag{5.36}$$

where ρ_0 and S_0 are the equilibrium values of ρ and S, ρ' and S' are amplitudes, and u is the propagation velocity along the x axis. Substituting (5.36) in (5.34) and (5.35),

$$\left[\left(\frac{u}{u_1} \right)^2 - 1 \right] \rho' + \left(\frac{\partial P}{\partial S} \right)_\rho \left(\frac{\partial \rho}{\partial P} \right)_s S' = 0 \tag{5.37}$$

$$\left(\frac{\partial T}{\partial \rho} \right)_s \left(\frac{\partial S}{\partial T} \right)_\rho \rho' + \left[\left(\frac{u}{u_2} \right)^2 - 1 \right] S' = 0 \tag{5.38}$$

where

$$u_1^2 = \left(\frac{\partial P}{\partial \rho}\right)_s$$

and

$$u_2^2 = \frac{\rho_s S^2}{\rho_n}\left(\frac{\partial T}{\partial S}\right)_\rho.$$

For (5.37) and (5.38) to be true simultaneously, the determinant of their coefficients must vanish, so that

$$\left[\left(\frac{u}{u_1}\right)^2 - 1\right]\left[\left(\frac{u}{u_2}\right)^2 - 1\right] = \left(\frac{\partial P}{\partial S}\right)_\rho\left(\frac{\partial \rho}{\partial P}\right)_s\left(\frac{\partial T}{\partial \rho}\right)_s\left(\frac{\partial S}{\partial T}\right)_\rho,$$

$$= \left(\frac{\partial P}{\partial T}\right)_\rho\left(\frac{\partial T}{\partial P}\right)_s. \tag{5.39}$$

By making use of standard thermodynamic identities it may readily be demonstrated that the product of partial derivatives on the right-hand side of (5.39) is equal to $(C_P - C_V)/C_P$. For HeII, the difference between C_P and C_V is very small so that, to an excellent approximation, the right-hand side of (5.39) is zero. Thus there are two possible solutions for the wave velocity

$$u^2 = u_1^2 = \left(\frac{\partial P}{\partial \rho}\right)_s \tag{5.40}$$

and

$$u^2 = u_2^2 = \frac{\rho_s S^2}{\rho_n}\left(\frac{\partial T}{\partial S}\right)_\rho = \frac{\rho_s}{\rho_n}\frac{TS^2}{C_V}. \tag{5.41}$$

If $u = u_1$, it is evident from (5.37) that S' must be zero (since there is no reason to suppose that either of the partial derivatives multiplying S' will be zero). The corresponding wave must therefore involve density oscillations at constant entropy, i.e. an ordinary sound wave. This is the mode known as *first sound*.

If $u = u_2$, a similar argument applied to (5.38) shows that $\rho' = 0$, corresponding to a wave in which entropy fluctuations occur at constant density. This mode is known as *second sound* (cf. §2.6 on second sound in solid helium, which is also a temperature wave but where no superfluid component is present). In second sound, in order to keep the overall density of the liquid constant, the normal and superfluid components must clearly move in antiphase with each other (Fig. 5.11b), in sharp contrast to ordinary (or first) sound where the two components move together (Fig. 5.11a).

First sound can be excited in the liquid in the usual ways; for example through the vibrations of a piezo-electric crystal which will move the normal and superfluid components in unison, as required. Detection of first sound can be accomplished by the inverse process, by use of a second piezo-electric

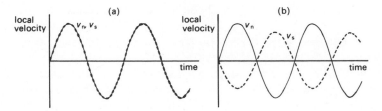

Figure 5.11 Differences between the fluid motion in first and the second sound. (*a*) In first sound, the normal and superfluid components (full and dashed curves, respectively) move *in phase* with each other such that $v_s = v_n$, and the temperature and entropy remain constant. (*b*) In second sound, the two components move *in antiphase* such that $\rho_s v_s + \rho_n v_n = 0$, and the overall density and pressure remain constant.

transducer. A convenient technique for the generation of second sound is to pass an a.c. current through a heater immersed in the liquid. The periodically-enhanced normal fluid density close to the heater propagates away from it as a continuous wave of second sound, at twice the frequency of the exciting current. Being a temperature wave, second sound is readily detected by means of a thermometer.

The measured velocities u_1 and u_2 are plotted as functions of temperature in Fig. 5.12. It will be noted that u_2 is very much smaller than u_1. Excellent agreement is obtained with the form of $u_2(T)$ predicted by (5.41) on the basis of values of ρ_s/ρ_n obtained by the Andronikashvili method (see §5.2). Indeed,

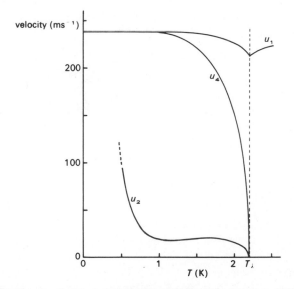

Figure 5.12 Wave mode velocities in bulk HeII, as functions of temperature T: first sound, u_1 (as tabulated by Donnelly, 1967); second sound, u_2 (after Peshkov, 1960); fourth sound, u_4 (Shapiro and Rudnick, 1965).

because the latter technique becomes rather inaccurate below 1.5 K, where ρ_n has become very small (Fig. 5.7b), it is now conventional to invert the argument and to derive values of $\rho_s(T)$ and $\rho_n(T)$ directly from $u_2(T)$ measurements, by use of (5.41) and (5.17).

Below c. 0.5 K, second sound no longer propagates as such. Thermal disturbances travel instead at the velocity of first sound. The reason is that the density of thermal excitations (see §5.4) is then insufficient for one to be able to speak realistically of a normal fluid component at all, and the hydrodynamic arguments leading eventually to (5.41) are therefore inapplicable. Under these conditions, a small heat pulse produces a group of virtually non-interacting phonons which travel ballistically at velocity u_1 between heater and detector.

The other wave mode which we consider in this section is *fourth sound*. Viscous damping normally prevents the transmission of sound through a liquid held in very narrow channels, for example in the interstices of a tightly packed powder. In the case of HeII, however, the (inviscid) superfluid can move quite freely under such conditions, so that a density wave can readily be transmitted. A first sound wave which encounters a superleak will consequently continue to propagate within it, but in an entirely different way. The normal fluid is effectively 'clamped' and it is to be expected that the characteristic propagation velocity will be quite different from either u_1 or u_2. Fourth sound clearly possesses some of the character of both first and second sound, however, in that there are periodic variations in both the overall density ρ and also in the ratio ρ_s/ρ_n. The magnitude of u_4 can be calculated from the equations of motion in much the same way as u_1 and u_2, except in that $\mathbf{v}_n$ is now set equal to zero. Again neglecting small terms, it can be shown (see for example, Donnelly, 1967) that

$$u_4^2 = (\rho_s/\rho)u_1^2 + (\rho_n/\rho)u_2^2 \tag{5.42}$$

where u_1 and u_2 are given, as before, by (5.40) and (5.41). By investigation of fourth-sound standing-wave resonances in HeII, excited by first-sound transducers in a powder-packed cavity, Shapiro and Rudnick (1965) were able to deduce the experimental values of u_4 shown in Fig. 5.12. These are in excellent agreement with (5.42), rising, as expected, from zero at T_λ and approaching u_1 at very low temperatures.

5.4 Superfluidity and excitations

The two-fluid model of HeII (§5.2) has been triumphantly vindicated by its success in correlating the observed properties of the liquid, and by its ability to predict in quantitative detail the occurrence of entirely new phenomena, such as the propagating thermal-wave modes (§5.3); it is, however, of no help in accounting for superfluidity itself. Indeed, the existence of a superfluid component constitutes the central premise of the model. In order to appreciate why the model works so well, and to gain some physical insight into the underlying reasons for the superfluid properties of HeII, we must now consider its nature at a rather more detailed level, including, in particular, the character of its low-lying thermally excited states. We will be describing the line of argument originally proposed by Landau (1941, 1947) and subsequently developed and justified by numerous other workers including, especially, Feynman (1955).

The thermal energy of an ideal Bose gas, to which we likened helium in §5.1, is comprised of the kinetic energies of the individual atoms described by the dispersion relation

$$E = p^2/2m_4 \qquad (5.43)$$

where p is the momentum of an atom and m_4 its mass. In making this comparison, however, it is essential always to bear in mind that HeII is not a gas but a liquid, so that the motion of any given atom must inevitably affect its neighbours through the interatomic forces. Landau argued that the thermal energy of HeII would therefore be likely to occur, not in the form of free-particle states described by (5.43), but rather as collective modes of the system as a whole, involving the coherent motion of large numbers of atoms. At low temperatures, where occupation numbers will be small and relaxation times correspondingly long, it is to be expected that these modes will be well defined. It is found experimentally that the heat capacity $C(T)$ of HeII varies as T^3 below 0.6 K (Fig. 5.13a), strongly suggesting that the modes in question may in fact be phonons, analogous to those in crystals (§2.3). Furthermore, in the same temperature range, the thermal conductivity $\kappa(T)$ is found to be roughly proportional to the diameter of the helium sample (Fig. 5.13b): behaviour which is strikingly reminiscent of the boundary scattering of phonons in

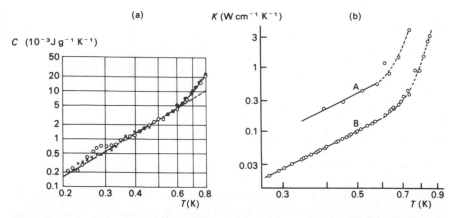

Figure 5.13 (a) The specific heat of HeII under its saturated vapour pressure (Wiebes et al., 1957). (b) The thermal conductivity of HeII under its saturated vapour pressure: A in a tube of 0.80 mm diameter; B in a tube of 0.29 mm diameter (Fairbank and Wilks, 1955). Note the logarithmic scales, in both cases.

crystals in the Knudsen limit (§2.5). The fact that C and κ both start to rise much more rapidly with T above 0.6 K may be interpreted as an indication that, in the case of HeII, some other type of excitation, in addition to phonons, becomes important at higher temperatures.

These general considerations have been amply confirmed through inelastic neutron-scattering experiments, which have enabled the dispersion relationship for the excitations in HeII to be mapped out in precise detail. It takes the form shown in Fig. 5.14 which, it is interesting to note, had been accurately predicted by Landau (apart from the plateau at high momentum) long before the definitive experiments had been performed. In practice, only those states corresponding to the two thickened regions of the curve become thermally populated to any significant extent. Close to the origin where, to a good approximation,

$$E = u_1 p \qquad (5.44)$$

the corresponding excitations may indeed be identified as longitudinal *phonons*, with the gradient u_1 being equal to the velocity of first sound. (In actuality, the low-energy phonon region exhibits very slight *anomalous dispersion*, in that $E(p)$ bends upwards to a marginal extent.) The local minimum in the curve is well described by

$$E = \Delta + (p - p_0)^2 / 2\mu_r. \qquad (5.45)$$

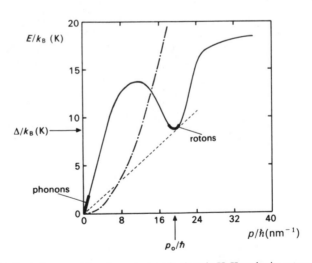

Figure 5.14 The dispersion curve for elementary excitations in HeII under its saturated vapour pressure at 1.1 K (Cowley and Woods, 1971). The energy E of an excitation (in temperature units) is plotted as a function of its momentum p (in units of reciprocal length). Only those states within the thickened regions of the curve, *phonons* and *rotons*, are populated to any significant extent in thermal equilibrium. The dashed line makes a tangent with the curve at a point which satisfies (5.50); and its gradient corresponds to v_L, the Landau critical velocity for roton creation (5.51). The dash-dot curve shows, for comparison, the dispersion relation (5.43) for a free ⁴He atom.

Table 5.1 Values of the roton parameters at very low temperatures in HeII (from Brooks and Donnelly, 1977).

Pressure (atm.)	0	25
Δ/k_B (K)	8.71	7.21
$p_0/\hbar$ (nm^{-1})	19.1	20.3
μ (m_4)	0.161	0.132

Excitations in this region are known as *rotons*. Their physical nature remains something of an enigma, although one possibility (Feynman, 1955) is that they may perhaps correspond to the quantum mechanical analogue of microscopic vortex rings (§5.5). The quantities Δ, p_0 and μ_r in (5.45) are known as roton parameters and represent respectively the energy, momentum and effective mass of a roton at the minimum. The roton parameters and velocity of sound depend on both pressure and temperature. Their low-temperature limiting values, given for two pressures in Table 5.1, remain accurate within 5% for temperatures up to about 1.5 K. Excitations in the region of the maximum are sometimes known as *maxons*. They occur only in negligible numbers in thermal equilibrium. By analogy with phonons in a crystal the excitations in HeII may be regarded as constituting a gas of quasiparticles, freely moving within an inert background at group velocities specified by the gradient at the relevant point on the dispersion curve.

It can be shown that the occurrence of superfluidity follows naturally if the thermal energy of the liquid occurs in the form of excitations which have the dispersion relationship shown in Fig. 5.14. We suppose that, as sketched in Fig. 5.15, a small object moves through HeII at $T = 0$ with initial velocity **v**. Then, if we assume for the moment that the creation of turbulence is rigorously forbidden in accordance with (5.16), the only way in which the object can lose kinetic energy, and thus slow down, is by creating excitations. We suppose that an excitation of energy E, momentum **p** is created and the reduced final velocity of the object is $\mathbf{v}_f$. In order that energy and momentum be conserved,

$$\tfrac{1}{2}mv^2 = \tfrac{1}{2}mv_f^2 + E \qquad (5.46)$$

$$m\mathbf{v} = m\mathbf{v}_f + \mathbf{p}. \qquad (5.47)$$

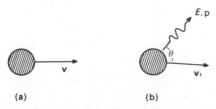

(a) (b)

Figure 5.15 An object which (a) travels through HeII with initial velocity **v** can reduce its kinetic energy by, as shown in (b), creating an excitation with energy E and momentum p. The object moves finally at a reduced velocity $\mathbf{v}_f$.

Eliminating $\mathbf{v}_f$,

$$\tfrac{1}{2}mv^2 = \tfrac{1}{2}m(v^2 - 2\mathbf{p}\cdot\mathbf{v} + p^2/m^2) + E$$

or

$$v\cos\theta = E/p + p/2m$$

where θ is, as shown, the angle between $\mathbf{v}$ and $\mathbf{p}$. Because $\cos\theta \leqslant 1$, we conclude that

$$v \geqslant E/p + p/2m \tag{5.48}$$

is a necessary condition for energy and momentum both to be conserved during the emission of an excitation of energy E, momentum $\mathbf{p}$. For a massive object the second term in (5.48) is negligible compared to the first, so that the condition reduces to $v \geqslant v_L$ where

$$v_L = (E/p)_{\min} \tag{5.49}$$

is known as the *Landau critical velocity*. Its value may be found by setting

$$\frac{d}{dp}\left(\frac{E}{p}\right) = 0$$

yielding

$$\frac{dE}{dp} = \frac{E}{p}. \tag{5.50}$$

Inspection of Fig. 5.14 reveals that, for HeII $(E/p)_{\min}$ lies in the region of the roton minimum; and (5.50) shows that v_L will in fact be specified by the gradient of a line drawn from the origin to make a tangent with the dispersion curve, contacting it at a point extremely close to the minimum itself as indicated by the dashed line. Thus, to a very good approximation,

$$v_L = \Delta/p_0 \tag{5.51}$$

(the exact value, found by substitution of (5.45) into (5.50) differing from Δ/p_0 by c. 1%). Inserting values of the roton parameters from Table 5.1, we find that v_L varies from $60\,\mathrm{m\,s}^{-1}$ at 0 bar to $46\,\mathrm{m\,s}^{-1}$ at 25 bar, just below the solidification pressure. Although we have derived (5.51) by considering the particular case of an object moving through the liquid, very similar arguments may also be applied to the situation of HeII flowing through a channel, leading to an identical conclusion.

The physical significance of v_L is that it represents a characteristic 'speed limit' for HeII: below v_L the liquid should retain its superfluid properties, but above v_L there is no longer any reason why dissipative effects should not occur. The validity of this picture may be tested by measuring the drag experienced by a moving object as a function of its velocity. Figure 5.16 plots the drag/velocity characteristic for a negative ion measured at a temperature low

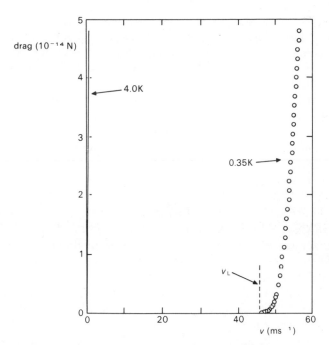

Figure 5.16   The drag force on a negative ion moving through HeII at 0.35 K, as a function of its speed $\bar{v}$ (Allum *et al.*, 1977). It is found that drag occurs only when $\bar{v}$ is equal to, or exceeds, the Landau critical velocity for roton creation, v_L. The behaviour of a negative ion in HeI at 4.0 K is also plotted, in order to emphasize the profound qualitative difference between the two cases.

enough for the residual gas of excitations to be neglected. As expected, the drag remains immeasurably small for velocities below a critical value, but then increases very rapidly for larger velocities. The observed critical velocity is in excellent numerical agreement with the value of v_L predicted by (5.51) using the roton parameters of Table 5.1. The drag/velocity characteristic for negative ions in liquid ^{4}He at 4.0 K is also shown in Fig. 5.16 in order to emphasize the profound qualitative difference in behaviour between HeI and HeII.

It is interesting to note that, on the basis of the above arguments, an ideal Bose gas would *not* be superfluid: for free particle dispersion, described by (5.43) and corresponding to the dash-dotted parabola in Fig. 5.14, it is clear that $(E/p)_{min}$ occurs at the origin and has the value zero. Thus there is no finite critical velocity below which dissipation cannot occur. It is also instructive to consider the manner in which the superfluidity of HeII disappears as the temperature rises towards T_λ. Neutron scattering experiments show that Δ decreases as $T \to T_\lambda$, implying a decrease of v_L. At the same time, however, the spectral linewidths increase rapidly owing to the decreased lifetime of the excitations, so that Δ and hence v_L become progressively less well defined: very close to T_λ the roton linewidth becomes comparable with Δ itself, corresponding to the effective 'washing out' of the precisely defined low temperature dispersion curve of Fig. 5.14 as the superfluid properties finally vanish.

The data of Fig. 5.16 amount to a convincing verification of Landau's explanation of superfluidity. In most physical situations, however, the measured critical velocities in HeII are found to be much smaller than v_L. This is because our assumption, that Fig. 5.14 includes all the permitted low-lying excited states of the liquid, is not entirely justified. While it is true that the indicated dispersion curve includes all permitted *elementary* excitations, it turns out that other (metastable) excitations involving quantized vortex rings and lines (§5.5) can also exist and that their creation is often characterized by critical velocities which are smaller than v_L: we discuss this point in more detail below (§5.7).

The excitation model of HeII may appear, at first sight, to be virtually the antithesis of the highly successful two-fluid model described in §5.2. Closer study reveals, however, that this is very far from being the case. Excitations in HeII carry momentum (in contrast to the quasi-momentum carried by phonons in crystals: see §2.4) and it can be shown that this corresponds to the momentum of the normal fluid component. Indeed, in a very real sense, the excitations *are* the normal fluid component. It is possible, not only to deduce the equations of two-fluid hydrodynamics from the excitation model, but also to use the measured shape of the dispersion curve for the excitations (Fig. 5.14) to predict (correctly) numerical values for the fundamental parameters of the two-fluid model, such as ρ_n. Thermodynamic quantities, such as C and S, may be calculated from the dispersion curve in the usual way, through application of statistical mechanics. The magnitude of the normal-fluid viscosity η_n may be accounted for in terms of interactions between the excitations (where, it should be noted, the Umklapp process of §2.5 cannot occur because there is no crystal lattice). A detailed discussion of the relationship between the two-fluid and excitation models of HeII may be found in, for example, Wilks (1967).

Fundamental theories of helium mostly seek to reproduce the dispersion curve by treating an assembly of hard-sphere bosons while taking appropriate account of the interactions between them. Once the dispersion curve has been derived successfully, superfluidity, the two-fluid model and the thermodynamic properties all follow in quantitative detail.

5.5 Quantized vortices in HeII

Up to this point, we have ignored any possibility of rotation in HeII. Indeed, (5.16) explicitly excludes any simple form of rotation, as can be seen from Stokes' theorem

$$\oint \mathbf{v} \cdot d\mathbf{l} = \int_A \nabla \wedge \mathbf{v} \cdot d\mathbf{A} \qquad (5.52)$$

which may be applied to any reducible closed circuit within the liquid. Thus, the *circulation* $\kappa_c = \oint \mathbf{v}_s \cdot d\mathbf{l}$ of the superfluid will be zero if $\nabla \wedge \mathbf{v}_s = 0$. At first sight, therefore, (5.16) appears to imply that the superfluid component within a rotating vessel of HeII will not go round at all, but will remain at rest. Careful measurements of the parabolic surface of HeII in a rotating vessel at different temperatures (Osborne, 1950) did not, however, support this conclusion. It seemed, rather, that the whole mass of helium rotated with the vessel like any ordinary liquid, in apparent violation of (5.16); which is, of course, an important equation of the two-fluid model. The paradox was resolved through the suggestion of Onsager (see London, 1954) that the superfluid circulation in HeII would be quantized: the liquid should be threaded with an array of quantized vortex lines, the net effect of which would be to simulate ordinary 'solid body' rotation.

The physical significance of this proposal is intimately related to the macroscopic wave function postulated in §1.6, which we will suppose to be applicable to the superfluid component of HeII (thus assuming, in effect, that the superfluid comprises the atoms occupying depletion levels as well as those of the condensate). We consider, first, the flow of superfluid around an annular channel between a pair of cylinders, as sketched in Fig. 5.17(a). We note that such a flow pattern does not in fact constitute a violation of (5.16), because the circuit is not a reducible one. We use (1.18) to calculate the circulation

$$\kappa_c = \oint \mathbf{v}_s \cdot d\mathbf{l} \tag{5.53}$$

$$= \frac{\hbar}{m_4} \oint \nabla S \cdot d\mathbf{l}$$

$$= \frac{\hbar}{m_4} \Delta S$$

where ΔS is the change in the phase of the wavefunction on passing round the circuit once. For a stationary state we require, as usual, that $\Delta S = n(2\pi)$ where

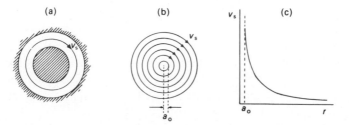

Figure 5.17 (a) Flow of the superfluid component with tangential velocity v_s around the annular space between two cylinders. (b) Sketch of a vortex line, with superfluid flowing around a core of radius a_0 at a velocity v_s, which depends inversely upon the radial distance r from the centre, as indicated in (c).

n is an integer, so that

$$\kappa_c = n(h/m_4).\qquad(5.54)$$

The circulation is therefore quantized in units of h/m_4. If we next imagine (Wilks, 1967) expanding the outer cylinder to infinity and gradually shrinking the inner one, a situation will eventually be reached (Fig. 5.17b) such that the inner cylinder can be removed altogether without affecting the flow pattern: this would then leave a cylindrical 'hole' in the liquid of radius a_0 such that the surface tension force inwards was exactly balanced by the Bernouilli (centrifugal) force outwards. Thus

$$\frac{2\gamma}{a_0} = \tfrac{1}{2}\rho v_s^2 \qquad(5.55)$$

where γ is the surface tension. If we carry out the integration of (5.53) round a circle of radius r, we find for the tangential superfluid velocity

$$v_s = \frac{\kappa_c}{2\pi r} \qquad(5.56)$$

whence, for $r = a_0$, (5.55) yields

$$a_0 = \frac{\rho\kappa_c^2}{16\pi^2\gamma}. \qquad(5.57)$$

Assuming $n = 1$ in (5.54), so that $\kappa_c = h/m_4$, and inserting the experimental values of γ and ρ, we find $a_0 \simeq 0.03$ nm. This result is, of course, of sub-atomic dimensions, where the concept of surface tension would be invalid. To obtain an accurate value of a_0, therefore, a more sophisticated (quantum-mechanical) calculation is needed. Nevertheless, these simple ideas have enabled us to deduce the principal features of a quantized vortex line in HeII. It consists of a cylindrical core of very small, perhaps sub-atomic, radius around which the superfluid circulates with a velocity which is inversely proportional to the distance from the core (Fig. 5.17c) in accordance with (5.56). Equation (5.16) will clearly be satisfied everywhere except on the core itself.

There is now ample experimental evidence showing that HeII inside a rotating container does in fact become threaded with an array of these vortex lines, arranged parallel to the axis of rotation, as sketched in Fig. 5.18(a). Each vortex core is subject to the combined flow fields of all the other vortex lines and it can be shown that, in consequence, the whole pattern (Fig. 5.18b) rotates with the container. The net effect is that the HeII will *seem* to rotate much like a solid body, provided one does not examine the liquid too closely, in accordance with the initial experimental observations.

For very small angular velocities of the container, however, such that only a few vortex lines are formed, the flow pattern will clearly deviate very considerably from that of solid-body rotation. For slow enough rotations of a container, it may not be energetically favourable to form even one vortex line.

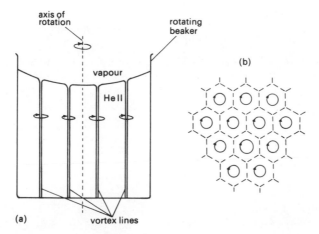

axis of rotation

rotating beaker

(b)

vapour

HeII

(a)

vortex lines

Figure 5.18 HeII in a rotating beaker. (*a*) Side view, with the vortex core diameters exaggerated for clarity. (*b*) Top view of a section of the surface: the dashed lines indicate points where the helium is at rest in a frame of reference rotating with the beaker; and the full arrows represent the superfluid velocity as viewed from the rotating frame.

If, under such conditions, rotating HeI is cooled through the lambda transition, the superfluid component will fall into a state of zero angular velocity (with respect to the fixed stars) as it is being formed. This astonishing process can actually occur, and it may be detected (Hess and Fairbank, 1967) from the consequent increase in the angular velocity of the freely suspended container, needed to ensure the overall conservation of angular momentum. The phenomenon is closely analogous to the Meissner effect (§4.3) but with rotation being 'squeezed' out of the helium rather than magnetic flux out of a superconductor.

Just as in the case of the flux lattice in a superconductor (Fig. 4.20*b*), an imaging technique has been developed which enables the vortices in a rotating vessel of HeII to be photographed (Yarmchuk *et al.*, 1979), with results as shown in Fig. 5.19, thereby providing a beautiful and convincing demonstration of this very remarkable quantum phenomenon.

Experimental values of the quantum of circulation, for comparison with (5.54), have been obtained by two main techniques. Vinen (1961) investigated the vibrations of a fine wire stretched in the liquid parallel to the axis of rotation. The presence of a stable superfluid flow field around the wire (which, in effect, replaced the core of the vortex discussed above) altered the vibrational modes in a characteristic way, enabling κ_c to be measured. Vinen's measurements (Fig. 5.20*a*) showed that κ_c was indeed quantized and that the magnitude of the quantum of circulation was equal to h/m_4 within experimental error.

Rayfield and Reif (1964) measured κ_c in a quite different way, by investigation of the energy–velocity relationship for quantized vortex rings. A

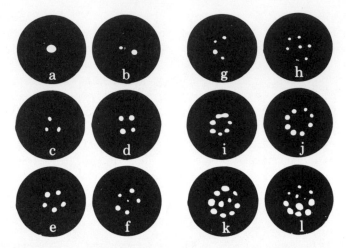

Figure 5.19 Stable (and metastable) vortex arrays in a rotating bucket of HeII for various angular velocities, photographed from above in the rotating frame of reference (Yarmchuk *et al.*, 1979). Each white spot corresponds to the presence of an individual quantized vortex line. The inner diameter of the bucket was 2 mm and its angular velocity varied between 0.3 and 0.9 rad s^{-1}. (Reproduced by courtesy of R.E. Packard.)

vortex ring in HeII consists simply of a vortex line, as described above, but which has been bent round and joined back on itself to form a toroid (inset to Fig. 5.20*b*). Each element of the ring is subject to the combined flow fields of all other parts of the ring, with the net effect that the ring moves forward with a characteristic velocity which depends primarily on κ_c and the radius R of the

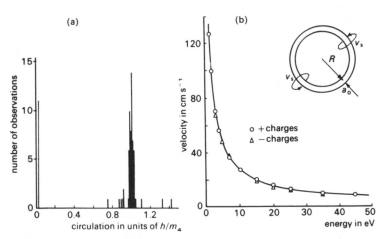

Figure 5.20 Determination of the quantum of circulation in HeII by two different methods. (*a*) Measurements of the circulation around a vibrating wire (Vinen, 1961) congregate predominantly about the value h/m_4. (*b*) The calculated variation of velocity with energy (full curve) for a charged vortex ring (inset) can be fitted extremely accurately to the experimental measurements (points) on the assumption that the quantum of circulation is h/m_4 (Rayfield and Reif, 1964).

ring (and only weakly on the core radius, a_0), and which *decreases* with increasing R. Such rings can be created in a charged state by means of positive or negative ions. Assuming that the rings carry only one quantum of circulation around their cores, their radii are then uniquely determined by the amount of energy they have been given. The measured velocity–energy relationship is shown by the points of Fig. 5.20(b). The (well-fitting) full curve is the calculated relationship for a core radius $a_0 = 0.12$ nm, on the assumption that $\kappa_c = h/m_4$.

These quite independent experiments clearly provide very strong support for (5.54) and may therefore be regarded as vindicating the hypothesis that the superfluid component of HeII is describable in terms of a macroscopic wavefunction.

5.6 The creeping superfluid film

Any solid surface dipping into HeII very quickly becomes covered with a film of liquid helium. Much the same behaviour occurs with other liquids, too, and for the same reason of straightforward physical adsorption. What is unusual about the HeII film, however, is that, in common with bulk HeII, it possesses a superfluid component. It is therefore able to flow across the surface of the substrate with quite remarkable rapidity. The effects of the creeping superfluid film were probably noted in some of the early experiments at Leiden, but they were not interpreted correctly until much later, as the result of work by (particularly) Rollin and Simon (1939) and Daunt and Mendelssohn (1939).

Any vessel which is partially immersed in a bath of HeII tends to fill up, via the film, until the inside and outside levels are equalized (Fig. 5.21a). Similarly, film flow will take place out of the vessel when the inside liquid level starts off

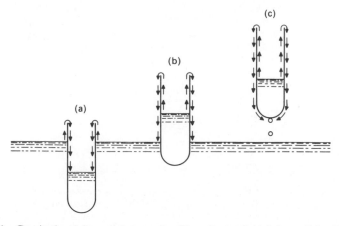

Figure 5.21   Gravitational flow of the creeping film of superfluid helium. (After Daunt and Mendelssohn, 1939.)

above that of the bath (Fig. 5.21*b*). The presence of the film can be demonstrated in a particularly dramatic way if the vessel is lifted clear of the helium bath (Fig. 5.21*c*). Droplets of liquid can then be seen collecting and falling back into the bath, until the vessel has completely emptied. The driving force in each case is gravity. The liquid is using the film, in effect, as a syphon as it 'attempts' to find its position of lowest attainable potential energy in the earth's gravitational field. It is found that the volume rate of flow is almost *independent of the magnitude of the driving force* provided by the level difference, but that it does depend on the minimum distance separating the highest point traversed by the film from a free liquid surface: in the experiment sketched in Fig. 5.21(*a*), for example, the vessel fills at a virtually constant rate until the inside and outside levels are equalized. When the levels do finally equalize in experiments such as those shown in Fig. 5.21(*a*) and (*b*), the kinetic energy of the film usually gives rise to some overshoot, followed by a weakly damped periodic variation of the levels usually known as the *Atkins oscillation*.

The equilibrium thickness of the film depends both on the nature (and cleanliness) of the substrate and on the height *h* above the free surface of bulk liquid. Typically, $d \sim 20$ nm at $h \sim 10$ mm. The thickness of an isothermal film is determined by a balance between the forces due to gravity and to the van der Waals' attraction to the substrate. We consider a semi-infinite flat wall in contact with bulk HeII, as sketched in Fig. 5.22, and we take as our zero of potential energy *U* that of an atom on the liquid surface at A, far beyond the range of the attractive interaction with the wall, so that $U_A = 0$. The potential energy of an atom on the surface of the film at B, above the curved part of the meniscus, will be

$$U_B = mgh - V(d)$$

where *V* is the potential energy due to the combined attraction of all the atoms

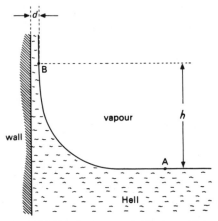

Figure 5.22   The free surface of HeII in isothermal equilibrium with its vapour and its film on a semi-infinite solid wall.

in the wall. A necessary condition for stability is, of course, that the potential energy of atoms on the helium surface should everywhere be the same. Thus $U_B = U_A = 0$ and

$$mgh - V(d) = 0. \tag{5.58}$$

On the assumptions that the van der Waals' potential between two isolated atoms separated by r can be written as

$$V(r) = Cr^{-6} \tag{5.59}$$

and by integration over all the atoms of the wall, it is straightforward to demonstrate that the net van der Waals' potential energy of the atom at B is $V(d) \propto d^{-3}$, whence (5.58) yields

$$d = \frac{k}{h^{1/3}} \tag{5.60}$$

Here, k is a constant which depends on the strength of the van der Waals' force between helium and the material of the wall. Measurements of the thickness of the film, based on a variety of techniques, are in reasonable accord with (5.60), although the agreement is seldom quantitatively exact.

It should perhaps be mentioned that, although the existence of the creeping film is readily enough demonstrated, quantitatively reproducible measurements of its properties are difficult, and call for very careful experimental design. In particular, the film thickness and the volume flow rate are critically dependent on the cleanliness and surface microstructure of the substrate and can be enormously increased by, for example, a thin layer of solidified air. Small heat fluxes can cause local evaporation of the film. Unintended film flow and, consequently, level differences can also be produced by small inadvertent temperature differences, via the thermomechanical effect (§5.2). In addition, it is expected that a moving film will be thinner than a static one under the same conditions.

Numerous experiments have also been carried out on *unsaturated* HeII films: that is, films which are not in equilibrium with a bath of bulk HeII and which can therefore be made arbitrarily thin. It is found that T_λ decreases below the bulk value with decreasing film thickness (a similar effect also being found for HeII held in very fine pores), but that superfluidity apparently persists right down to thicknesses corresponding to about two liquid layers of helium atoms. There is evidence that the first one or two layers may, in fact, be solid (that is, non-mobile) as a result of the very strong van der Waals' attraction so close to the substrate.

The narrowness of the 'channel' provided even by saturated films ensures that the normal fluid component remains effectively clamped to the substrate. The superfluid component, by contrast, is able to flow entirely without hindrance (provided that it does not exceed a critical velocity: see §5.7). In addition to the steady-state superflow described above, surface waves in which the superfluid oscillates parallel to the substrate are also possible, and are known as *third sound*. Because only the superfluid moves, there is consequently a periodic variation in the thickness of the film, with corresponding fluctuations in its temperature: the thickened regions, where there is excess superfluid, are transiently at a lower temperature and *vice versa* (although the amplitude of the temperature fluctuations is, to some extent, reduced by evaporation/condensation of helium atoms from hotter to colder regions).

A detailed treatment of third sound involves the solution of the equations of motion subject to appropriate boundary conditions (Atkins and Rudnick, 1970), much as in the cases of first, second and fourth sound (§5.3). The phase velocity u_3 of third sound can, however, be estimated by modifying the ordinary expression for the velocity v_w of a surface wave on a shallow classical liquid of depth d

$$v_w^2 = \left(\frac{f\lambda}{2\pi} + \frac{2\pi\gamma}{\rho\lambda} \right) \tanh \frac{2\pi d}{\lambda}. \qquad (5.61)$$

Here, f is the attractive force of the substrate per unit mass of liquid, γ the surface tension, ρ the density and λ the wavelength. For long wavelengths, where the second term is negligible compared to the first and the tanh approaches its argument, (5.61) becomes

$$u_3^2 = \rho_s f d / \rho \qquad (5.62)$$

where the factor ρ_s/ρ has been inserted to take account of the fact that only the superfluid moves. Third sound can be excited and detected optically, thermally or (for very low frequencies) through the use of standpipes. Velocities lie typically in the range $0.5\text{--}40\,\mathrm{ms}^{-1}$.

An excellent survey of experimental and theoretical work on almost all aspects of the HeII film will be found in the review article by Brewer (1978).

5.7 Critical velocities

Many of the properties of HeII are characterized by (one or more) *critical velocities* at which abrupt changes of behaviour take place. Typically, the superfluid component behaves as an ideal (inviscid) fluid below the critical velocity, but exhibits dissipative behaviour at higher velocities. One example which we have already considered (in §5.4) is the sudden onset of drag on a negative ion moving through the liquid (Fig. 5.16) corresponding to the creation of rotons above the Landau critical velocity v_L.

Most critical velocities in HeII are a good deal smaller than v_L, however, and are quantitatively much less well understood, even though they may be equally dramatic in form. For example, the torque on a propeller mounted in a superfluid wind tunnel (a tube where *only* the superfluid moves) is found (Craig and Pellam, 1957) to remain zero up to about $6\,\mathrm{mm\,s}^{-1}$, but to rise quite

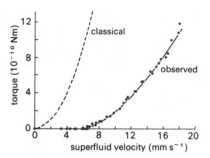

Figure 5.23   The measured torque on a propeller suspended in a flowing stream of the superfluid component of HeII, as a function of the flow velocity (Craig and Pellam, 1957).

rapidly for larger velocities as shown in Fig. 5.23, again quite contrary to the behaviour expected in a classical fluid. The general form of this graph is rather similar to Fig. 5.16, but the relevant critical velocity is here several hundred times smaller. Another example of a critical velocity may be observed (Keller and Hammel, 1966) by measuring the pressure gradient ∇P necessary to force HeII through a fine channel. Flow can apparently occur under a zero pressure gradient, but only up to a critical velocity of a few tens of mm s^{-1}. For larger velocities, a finite value of ∇P is required to maintain the flow.

Critical velocities are also involved in the flow of the creeping superfluid film (§5.6). In the experiments sketched in Fig. 5.21, the net transfer rate through the film will be limited to the rate at which helium can traverse the lip of the beaker, where the film will be at its thinnest, in accordance with (5.60). In (a) the film thickness at this point will, of course, remain constant because the height of the lip above the highest level of bulk liquid does not change. Consequently, if we postulate that there exists some critical velocity below which dissipation is entirely absent, but above which dissipation increases very rapidly, then we would expect to see a constant transfer rate, independent of level difference, exactly as observed experimentally.

The flow of heat in bulk HeII is characterized by critical velocities. The simplest type of experiment is based on the arrangement sketched in Fig. 5.10(a), where a normal fluid/superfluid counterflow takes place in a narrow tube. For small heat fluxes, corresponding to small normal and superfluid flow velocities, the temperature difference ΔT is directly proportional to the heat input $\dot{Q}$, and values of η_n can be deduced as previously described. At some larger heat flux, however, ΔT suddenly starts to rise much more rapidly with $\dot{Q}$. Although the transition often exhibits some hysteresis, it is possible to define a critical heat flux W_c and hence to calculate the corresponding critical counterflow velocity.

In each of these different types of flow experiments, the magnitude v_c of the critical velocity is found to be dependent on the dimensions of the channel used. The value of v_c varies from, for example, 0.3 mm s^{-1} in a 4 mm channel up to $10\,ms^{-1}$ for flow through 20 nm holes etched in irradiated mica. There is overwhelming evidence to show that critical velocities in HeII (apart, of course, from the Landau critical velocity v_L) correspond to a transition of the liquid to a *turbulent state*. In the thermal counterflow experiments, this consists of a more or less randomly tangled mass of the quantized vortex lines discussed in §5.5. The vortex cores are envisaged (in most situations) as taking up a structure rather like a tangle of steel wool or of spaghetti. Of course, each section of core is subject to the flow fields around all the others, so that the whole tangle is in a state of constant motion: a can of worms might in some ways be a more apt analogy, therefore (although it must be borne in mind that the vortex lines cannot have free ends, which would be unstable). Although the vortex lines represent quantized motions of the superfluid, their cores effectively scatter the thermal excitations which constitute the normal fluid. The presence of vortices leads, therefore, to a mutual friction between the normal and superfluid components, and hence to the dissipative effects which are observed experimentally. Measurements of the attenuation coefficient of second sound (§5.3) can thus be used to provide a particularly sensitive method of detecting the presence of vortex lines in HeII (Hall and Vinen, 1956).

The nature of the turbulent state was originally elucidated by Vinen (1961). He proposed that the vortex tangle was homogeneous and isotropic, and that its equilibrium density was determined by a balance between line creation processes (driven by the relative velocity of the two fluids) and annihilation processes (which could occur if, for example, two sections of line with opposite circulation approached each other closely enough). An excellent overview of current experimental and theoretical understanding of the turbulent state of HeII will be found in the review by Tough (1982).

Notwithstanding the considerable progress which has been made in understanding the growth, equilibrium and decay of superfluid turbulence, the mechanisms by which the vortex lines or rings are created in the first place remain something of an enigma. The observed decrease in critical velocity v_c with increasing channel size is actually to be expected (Atkins, 1959) if one postulates that the onset of dissipation corresponds to the formation of a vortex ring, and one applies a generalized Landau argument (§5.4) to estimate the relevant critical velocity. The agreement obtained between experiment and theory, however, is more qualitative than quantitative. The most likely explanation for the lack of agreement is that the values of v_c measured in bulk flow experiments almost always correspond to the velocities needed to expand vortex lines *already present* in the liquid and not to the velocities needed to create them *ab initio*. In fact, it is highly probable that studies of vortex nucleation by negative ions (Bowley *et al.*, 1982, and references therein) represent the only type of experiment on bulk HeII in which intrinsic critical velocities, corresponding to the actual creation of vortices, have been reliably observed: the ions are so small (radii $\simeq 1$ nm) that the influence of remanent vorticity is expected to be negligible.

A mechanism to account for what appear to be intrinsic critical velocities was proposed by Iordanskii (1965) and Langer and Fisher (1967). Their suggestion, essentially, was that dissipation in HeII is initiated through the formation of vortices as the result of thermal fluctuations. On this model, critical velocities cannot be defined in an absolute sense. Rather, the transition rate is predicted to be an exceedingly fast function of velocity, so that v_c represents, in effect, that velocity at which the reciprocal of the rate becomes comparable with the measurement time of the experiment. The theory has had a considerable measure of success, particularly in experiments above c. 1 K (Langer and Reppy, 1970); but, for lower temperatures, it generally seems to predict dissipation rates which are much smaller than those measured.

5.8 Other boson fluids

Given that the extraordinarily rich range of phenomena exhibited by liquid ^{4}He below T_λ is apparently associated with Bose–Einstein condensation, it is only natural to seek other boson systems which might be expected to display behaviour of comparable interest. They are hard to find; although there is, of course, a sense in which Cooper-paired fermions in superconductors (§4.5) or superfluid ^{3}He (§6.4) can be regarded as undergoing a Bose–Einstein condensation. As noted in §1.5, helium is the only form of bulk matter which has not already become solid at the temperature where a Bose–Einstein condensation might otherwise have been anticipated. Photons and phonons both constitute boson assemblies but, because there is no requirement for conservation of quasiparticles, the arguments of §5.1 leading to condensation do not apply.

A ^{4}He *monolayer* on an atomically flat substrate constitutes a 2-dimensional boson assembly of very considerable interest (Dash and Schick, 1978), but a Bose–Einstein condensation is not expected and neither does it appear to have been seen experimentally. (It is straightforward to show that arguments analogous to those of §5.1, but based on the density of states appropriate to a 2-D assembly in place of (5.5), do not imply a condensation.) Superfluidity in 2-D is still to be expected (Kosterlitz and Thouless, 1973), however, with some distinctive features beautifully verified in subsequent experiments (Bishop and Reppy, 1978).

The only other system in which there appears to be a reasonable chance that a Bose–Einstein condensation may eventually be observed is hydrogen. In equilibrium, hydrogen forms tightly bound diatomic molecules in which the electron spin angular momenta are antiparallel; and, at low temperatures, hydrogen usually forms a molecular solid. The force between a pair of hydrogen atoms whose spins are parallel is weakly *repulsive*, however, and so spin-polarized hydrogen may be expected to remain gaseous to arbitrarily low temperatures. Bose–Einstein condensation is therefore to be anticipated for low enough temperatures, or high enough densities, in accordance with (5.12).

The main experimental problem is, of course, to stabilize the polarized hydrogen, given the huge energy release which occurs when an atom undergoes a spin-reversal and combines with another to form a molecule. Considerable progress has now been made towards this end, however, particularly through use of a combination of powerful magnetic fields and low temperatures, and by coating all surfaces with a film of HeII in order to separate the hydrogen from magnetic impurities in the walls (which can catalyse molecular recombination). Details will be found in the review article by Silvera (1982). It is also just possible that it may prove feasible to supercool ordinary molecular *liquid* hydrogen to below the temperature where a Bose–Einstein condensation may be anticipated.

Bibliography

References

Allen, J.F. and Jones, H. *Nature* **141**, 243 (1938).
Allum, D.R., McClintock, P.V.E., Phillips, A. and Bowley, R.M. *Phil. Trans. Roy. Soc. (Lond.)* **A284**, 179 (1977).
Andronikashvili, E. *J. Phys. (USSR)* **10**, 201 (1946); translated in Z.M. Galasiewicz (1971), p. 154 (see below).
Atkins, K.R. and Rudnick, I., in *Progress in Low Temperature Physics*. Ed. C.J. Gorter, vol. VI, ch. 2, North-Holland, Amsterdam (1970).
Bishop, D.J. and Reppy, J.D. *Phys. Rev. Lett.* **40**, 1727 (1978).
Bowley, R.M., McClintock, P.V.E., Moss, F.E., Nancolas, G.G. and Stamp, P.C.E. *Phil. Trans. Roy. Soc. (Lond.)* **A307**, 201 (1982).
Brewer, D.F., in *The Physics of Liquid and Solid Helium*. Ed. K.H. Bennemann and J.B. Ketterson, part II, p. 573 (1978: see below).
Brewer, D.F. and Edwards, D.O. *Proc. Roy. Soc. (Lond.)* **A251**, 247 (1959).
Brooks, J.S. and Donnelly, R.J. *J. Phys. Chem. Ref. Data* **6**, 51 (1977).
Campbell, L.J. *Phys. Rev.* **B27**, 1913 (1983).
Cowley, R.A. and Woods, A.D.B. *Can. J. Phys.* **49**, 177 (1971).
Craig, P.P. and Pellam, J.R. *Phys. Rev.* **108**, 1109 (1957).
Dash, J.G. and Schick, M., in *The Physics of Liquid and Solid Helium*. Eds. K.H. Bennemann and J.B. Ketterson, part II, p. 497 (1978: see below).
Daunt, J.G. and Mendelssohn, K. *Nature* **143**, 719 (1939); and *Proc. Roy. Soc. (Lond.)* **A170**, 423, 439 (1939).
Fairbank, H.A. and Wilks, J. *Proc. Roy. Soc. (Lond.)* **A231**, 545 (1955).
Feynman, R.P., in *Progress in Low Temperature Physics*. Ed. C.J. Gorter, vol. I, ch. 2; North-Holland, Amsterdam (1955).
Hall, H.E. *Proc. Phys. Soc.* **A67**, 485 (1954).
Hall, H.E. and Vinen, W.E. *Proc. Roy. Soc. (Lond.)* **A238**, 204 (1956).
Hess, G.B. and Fairbank, W.M. *Phys. Rev. Lett.* **19**, 216 (1967).
Iordanskii, S.V. *Zh. Eksp. Teor. Fiz. (USSR)* **48**, 708 (1965); translated in *Soviet Phys. J.E.T.P.* **21**, 467 (1965).
Kapitza, P.L. *J. Phys. (USSR)* **4**, 181 (1941); translated in ed. Z.M. Galasiewicz (1971), p. 114 (see below).
Keller, W.E. and Hammel, E.F. *Physics* **2**, 221 (1966).
Kosterlitz, J.M. and Thouless, D.J. *J. Phys. C: Solid St. Phys.* **6**, 1181 (1973).
Landau, L.D. *J. Phys. (USSR)* **5**, 71 (1941), and *J. Phys. USSR* **11**, 91 (1947); both papers translated in Z.M. Galasiewicz (1971), pp. 191 and 243 (see below).
Langer, J.S. and Fisher, M.E. *Phys. Rev. Lett.* **19**, 560 (1967).
Langer, J.S. and Reppy, J.D., in *Progress in Low Temperature Physics*. Ed. C.J. Gorter, vol. VI, ch. 1; North-Holland, Amsterdam (1970).

London, F. *Nature* **141**, 643 (1938).
Osborne, D.V. *Proc. Phys. Soc.* **63**, 909 (1950).
Peshkov, V.P. *Zh. Eksp. Teor. Fiz.* (USSR), **38**, 799 (1960); translated in *Soviet Phys. J.E.T.P.* **11**, 580 (1960).
Rayfield, G.W. and Reif, F. *Phys. Rev.* **136**, A1194 (1964).
Rollin, B.V. and Simon, F. *Physica* **6**, 219 (1939).
Sears, V.F., Svensson, E.C., Martel, P. and Woods, A.D.B. *Phys. Rev. Lett.* **49**, 279 (1982).
Silvera, I.F. *Physica B + C* **109** and **110**, 1499 (1982).
Shapiro, K.A. and Rudnick, I. *Phys. Rev.* **137**, A1383 (1965).
Tough, J.T., in *Progress in Low Temperature Physics*. Ed. D.F. Brewer, vol. VIII, ch. 3; North-Holland, Amsterdam (1982).
Vinen, W.F., in *Progress in Low Temperature Physics*. Ed C.J. Gorter, vol. III, ch. 1; North-Holland, Amsterdam (1961).
Wiebes, J., Niels-Hakkenberg, C.G. and Kramers, H.C. *Physica* **23**, 625 (1957).
Yarmchuk, E.J., Gordon, M.J.V. and Packard, R.E. *Phys. Rev. Lett.* **43**, 214 (1979).

Further reading

Ahlers, G. 'Experiments near the superfluid transition in ^{4}He and ^{3}He $-$ ^{4}He mixtures', p. 85ff in Bennemann and Ketterson (1976), q.v.
Atkins, K.R. *Liquid Helium*. Cambridge University Press, Cambridge (1959).
Bennemann, K.H. and Ketterson, J.B. (eds.) *The Physics of Liquid and Solid Helium*. Wiley, New York, part I (1976) and part II (1978).
Donnelly, R.J. *Experimental Superfluidity*. University of Chicago Press, Chicago (1967).
Galasiewicz, Z.M. (ed.) *Helium 4*. Pergamon (1971). (Reprints of a selection of classic papers on ^{4}He, with an introduction by the editor).
Gorter, C.J. (ed.) *Progress in Low Temperature Physics*. North-Holland, Amsterdam (1955–). (A series: recent volumes ed. by Brewer, D.F.).
Hallock, R.B. 'Resource Letter SH-1: superfluid helium', *Amer. J. Phys.* **50**, 202 (1982).
Keller, W.F. *Helium-3 and Helium-4*, Plenum, New York (1969).
Lane, C.T. *Superfluid Physics*. McGraw-Hill, New York (1962).
London, F. *Superfluids*. Vol. II (Superfluid Helium), Wiley, New York (1954); reprinted Dover, New York (1964).
Mandl, F. *Statistical Physics*. Wiley, London (1971).
Putterman, S.J. *Superfluid Hydrodynamics*. North-Holland, Amsterdam (1974).
Tilley, D.R. and Tilley, J. *Superfluidity and Superconductivity*. Van Nostrand Reinhold, New York (1974); and new edition (Adam Hilger, forthcoming).
Wilks, J. *The Properties of Liquid and Solid Helium*. Clarendon Press, Oxford (1967); and an abridged and updated version *An Introduction to Liquid Helium*, Clarendon Press (1970).

Ciné films

Allen, J.F. and Armitage, J.G.M. *Superfluid Helium*. (1982, 5th edition, 43 min., 16 mm, colour, sound); available from the producers at: Department of Physics, University of St. Andrews, St. Andrews, KY16 9SS, Scotland.
Rudnick, I. *The Unusual Properties of Liquid Helium*. (1977, 17 min., 16 mm, colour, sound); available from the producer at: Department of Physics, University of California, Los Angeles, California 90024, USA.

6 Liquid helium-3 and ^{3}He–^{4}He solutions

6.1 Influence of Fermi–Dirac statistics

Liquid ^{3}He, because of its lower atomic mass, has a larger zero-point energy (§1.5) than liquid ^{4}He. In consequence, its atoms are more widely spaced, on average, so that the analogy with a gas is likely to be even better than in the case of liquid ^{4}He. Unlike ^{4}He, however, the ^{3}He atom (2 protons, 1 neutron, 2 electrons) is composed of an uneven number of fundamental particles. It therefore possesses a resultant spin, of $\frac{1}{2}\hbar$. The ^{3}He atom is thus a *fermion*, with anti-symmetric wavefunction; and an assembly of ^{3}He atoms is consequently subject to the Pauli exclusion principle in much the same way as, for example, electrons in atoms or the quasi-free electron gas in a normal metal.

We suppose, therefore, that liquid ^{3}He may be regarded, at least to a first approximation, as an ideal Fermi gas. Thus, at very low temperatures ($T \ll T_F$, where $T_F = E_F/k_B$ is the Fermi temperature), there will be a sharply defined Fermi sphere of occupied states in momentum spacce: a picture that could hardly be in more striking contrast to the macroscopically occupied zero-momentum ground state (condensate) of liquid ^{4}He discussed in §5.1. It is to be expected, therefore, that the low-temperature properties of liquid ^{3}He will be profoundly different from those of liquid ^{4}He and, as shown below, there is ample experimental evidence that this is indeed the case. The large qualitative differences between liquid ^{3}He and liquid ^{4}He that are observed would hardly be likely to result purely from the different atomic mass, so that they may be regarded as a direct and dramatic demonstration of the fundamental importance of the underlying quantum statistics.

It must be emphasized that the transition from high temperature (classical) behaviour to the low temperature (quantal) regime occurs in an entirely different way for boson and fermion assemblies. For bosons, a phase transition occurs at the condensation temperature (§5.1) whereas, for fermions, the transition takes place gradually and can be regarded as complete only when the temperature has fallen below $c.\ 0.1\ T_F$.

Just as in the case of ^{4}He, interatomic forces cause the behaviour of the actual liquid to depart from that predicted for an ideal gas. For some purposes, the resultant modifications in properties can be taken account of by regarding ^{3}He atoms in the liquid as being replaced by 'quasiparticles', having an effective mass m_3^* which is larger than the mass m_3 of a free ^{3}He atom. There are many instances, however, where this approach is insufficient by itself and it is then necessary to use Landau's *Fermi liquid theory*, as outlined in §6.3. An important consequence of the interactions between the atoms is that T_F for liquid ^{3}He is considerably smaller than the value (c. 5 K) calculated on the basis of the bare atomic mass for an ideal gas of the same density.

6.2 Properties of normal liquid ^{3}He

Above 1 K, liquid ^{3}He behaves much like a dense classical gas and its behaviour is similar to that of HeI. As the liquid is cooled to lower temperatures, however, its properties undergo a gradual change until it has entered a regime of profoundly different behaviour extending from c. 50 mK down to the superfluid transition. We describe below some selected properties and compare them, in each case, with those expected for an ideal Fermi gas.

The heat capacity of liquid ^{3}He is plotted as a function of temperature in Fig. 6.1. For the normal liquid (a, and b for $T > T_c$) $C(T)$ is entirely different from that of liquid ^{4}He (cf. Fig. 5.1), but closely resembles the behaviour expected of the ideal gas. In particular, it tends towards a linear dependence on T at very low temperatures, much like $C(T)$ for the electron gas in a normal metal (§3.2). On the assumption of an ideal Fermi gas, the limiting low-temperature behaviour of the molar specific heat (in analogy with equation 3.19) should be given by

$$C = \pi^2 k_B^2 N T / 2 E_F \qquad (6.1)$$

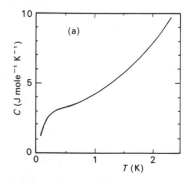
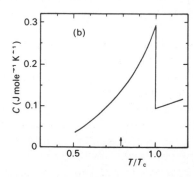

Figure 6.1   The specific heat C of liquid ^{3}He as a function of temperature T: (a) at relatively high temperatures under the saturated vapour pressure (after Wilks, 1967); (b) in the vicinity of the superfluid transition temperature $T_c = 2.8$ mK at the melting pressure (after Halperin et al., 1976), where the position of the ^{3}He-A to ^{3}He-B transition is marked by the arrow.

which, on insertion of the relevant constants for liquid ^{3}He at the SVP, yields $C = 1.00\,RT$. Experimental values, however, are about three times larger than this prediction. The discrepancy can be accommodated if, as indicated in the previous section, the m_3 which enters E_F is replaced by an effective mass m_3^* which is some three times larger than the bare atomic mass.

The viscosity of liquid ^{3}He rises rapidly with decreasing temperature below 1 K (Fig. 6.2), in particularly striking contrast to the behaviour of liquid ^{4}He. For an ideal gas, the viscosity is given by simple kinetic theory as

$$\eta = \tfrac{1}{3}\rho\tau v^2 \qquad (6.2)$$

where τ is the average time between collisions and v is the mean atomic speed. In the particular case of a degenerate Fermi assembly, energetically accessible empty states into which the colliding atoms can scatter exist only within a narrow energy range of width $c.\ k_B T$ centred on the Fermi surface. Only those atoms which are very close to the Fermi surface can undergo scattering, and the relevant speed to use in (6.2) will therefore be the Fermi velocity v_F. The temperature-dependence of τ can be deduced by noting that τ must be inversely proportional to the probability of finding *two* empty states: one for each of the scattered particles. The probability of finding one empty state is proportional to $k_B T$. Thus τ will be inversely proportional to $(k_B T)^2$ and, because $v = v_F$ is temperature-independent for $T \ll T_F$, (6.2) yields

$$\eta \propto T^{-2}. \qquad (6.3)$$

This result is in reasonable agreement with the limiting low-temperature behaviour observed experimentally down as far as the superfluid transition.

The thermal conductivity κ of liquid ^{3}He passes through a minimum near 0.2 K and then rises (Fig. 6.3) with decreasing temperature. (The data are plotted in terms of the *magnetic temperature* T^*, measured with a paramagnetic susceptibility thermometer (§7.5), which differs very slightly from the Kelvin temperature.) Again, this behaviour may readily be understood by analogy

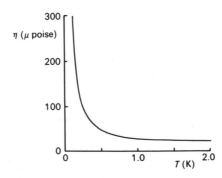

Figure 6.2 The viscosity η of normal liquid helium as a function of temperature T. (After Wilks, 1967.)

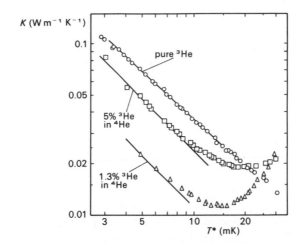

Figure 6.3   The thermal conductivity κ of pure liquid ^{3}He and of two liquid ^{3}He–^{4}He solutions as functions of the magnetic temperature T^*. (After Wheatley, 1970.)

with that expected for an ideal Fermi gas. Simple kinetic theory gives

$$\kappa = \tfrac{1}{3}\rho C v^2 \tau \tag{6.4}$$

where C is the specific heat per unit mass and, as before, τ is the mean time between atomic collisions. For $T \ll T_F$, we may argue once more that $v \simeq v_F$ and $\tau \propto (k_B T)^{-2}$. Since $C \propto T$ (see above), we may deduce from (6.4) that

$$\kappa \propto T^{-1} \tag{6.5}$$

which, again, is in reasonable accord with experimental observation.

In marked contrast with liquid ^{4}He, liquid ^{3}He is a magnetically active material, as a result of the unpaired spins in its nuclei. Although the magnetism is relatively weak because of the small size of the nuclear magneton (5.4×10^{-4} of the Bohr magneton characterizing electronic magnetism), the magnetic properties of ^{3}He are extremely important. They are of considerable fundamental interest in their own right and, in addition, have played a crucial role in the experimental investigation of the superfluid phases (§6.4) of the liquid. The existence of the nuclear spin is also directly responsible for the minimum in the melting curve (see below).

The measured paramagnetic susceptibility χ of liquid ^{3}He is shown in Fig. 6.4, plotted in the form of χT against T. Although χ varies linearly with T^{-1} in accordance with Curie's law above 1 K, it becomes almost temperature-independent at very low temperatures. It may readily be demonstrated that this is precisely the behaviour to be expected of an ideal Fermi gas. The nuclear dipoles will all 'align' either parallel or antiparallel to an applied magnetic field B, thereby losing or gaining energy $\mu_N B$, where μ_N is the nuclear magneton. The situation is illustrated schematically in Fig. 6.5(a)

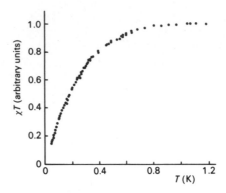

Figure 6.4 The magnetic susceptibility χ of liquid ^{3}He plotted as χT (in arbitrary units) against the temperature T. (After Thomson, Meyer and Adams, 1962.)

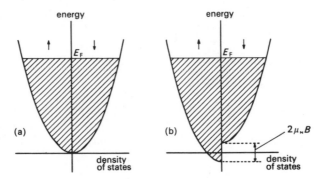

Figure 6.5 Schematic diagram illustrating the origin of the temperature-independent Pauli paramagnetism of normal liquid ^{3}He in the degenerate limit. For explanation, see text.

and (b), where the total (kinetic and magnetic) energies for ^{3}He atoms in parallel and antiparallel states are plotted against the density of states respectively on the left- and right-hand sides of each diagram. In (a), there is no applied field, and the energy is independent of spin orientation. In (b), the field has been applied, raising the energies of the antiparallel spins by $\mu_N B$ and reducing those of the parallel spins by the same amount. Following a transient unstable configuration, the energies of the two Fermi surfaces equalize as the result of transitions from the antiparallel to the parallel state. When equilibrium has been restored, there will be ΔN more parallel than antiparallel spins and, consequently, a net magnetic moment equal to $\mu_N \Delta N$. If the density of states at the Fermi surface is $D(0)$, then $\Delta N = \mu_N B D(0)$ and the paramagnetic susceptibility per unit volume

$$\chi = \mu_0 \mu_N \Delta N/B = \mu_0 \mu_N^2 D(0) \qquad (6.6)$$

Thus, χ will be independent of temperature (provided that $T \ll T_F$ so as to produce well-defined Fermi surfaces, as sketched) in accord with the measurements (Fig. 6.4). This result is, of course, exactly paralleled by the well-known Pauli paramagnetism of the electron gas in a normal metal.

The Clausius–Clapeyron equation (see, for example, Zemansky and Dittman, 1981)

$$\left(\frac{dP}{dT}\right)_m = \frac{\Delta S_m}{\Delta V_m} \tag{6.7}$$

carries interesting implications for ³He. The m subscript refers to the melting curve and ΔS and ΔV are the changes in entropy and volume which occur on melting. The existence of the minimum in the melting curve (Figs. 1.9b and 6.6b), giving a negative $(dP/dT)_m$ below 0.3 K, shows, because the volume increases on melting, that the entropy of the solid phase of ³He must be *greater* than that of the liquid in this temperature range, which is of course a complete reversal of the normal state of affairs. This curious situation may readily be accounted for if it is noted that ³He atoms in the solid are (fairly well) localized on lattice sites and are consequently described not by Fermi, but by Boltzmann statistics. The entropy of the solid in zero magnetic field below 1 K arises almost entirely from the nuclear spins which (except at the lowest

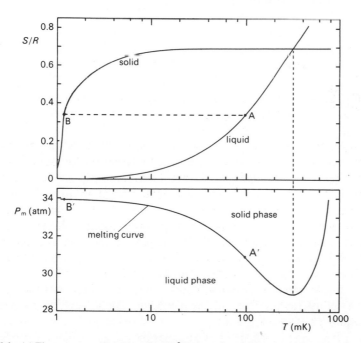

Figure 6.6 (a) The entropy of liquid and solid ³He in units of the gas constant R, as functions of temperature T. (b) The melting pressure P_m of ³He as a function of T. (From values tabulated by Betts, 1976.)

temperatures) are randomly distributed between two permitted states. The molar entropy of the solid will consequently be equal to $R \ln 2$ in accordance with the Boltzmann–Planck equation (1.2), and thus independent of temperature. The molar entropy of the liquid may be found in the usual way by integrating the specific heat:

$$S = \int (C/T) \, dT$$

where C is now the specific heat per mole of ^{3}He. We have already seen that the specific heat varies approximately linearly with T. The liquid entropy will also, therefore, decrease approximately linearly with decreasing T so that it must, eventually, fall below that of the solid. Experimental values (deduced from $C(T)$ measurements) of the two entropies are plotted in Fig. 6.6a where it can be seen that the two curves do, indeed, cross each other at exactly the same temperature as that of the minimum in the melting curve (Fig. 6.6b), in excellent agreement with (6.7). The sudden descent towards zero of the solid entropy at very low temperatures (in accordance with the Third Law) corresponds to an antiferromagnetic ordering transition at about 1 mK.

6.3 The Landau theory of liquid ^{3}He

It will be clear from the foregoing discussion that the behaviour of liquid ^{3}He is qualitatively very similar to that predicted for an ideal Fermi gas; a quantitative comparison reveals, however, that there are large numerical discrepancies. That such differences should exist is, of course, hardly surprising given that the interatomic forces, although weak, are certainly far from negligible: the assembly is, after all, a liquid. The differences cannot be accounted for satisfactorily simply in terms of an effective mass m_3^*, because a choice of m_3^* so as to match the ideal Fermi gas model to one particular physical property will, in general, not be appropriate for other properties. Landau's (1956) *Fermi liquid theory*, however, takes explicit account both of the Fermi statistics and of the interatomic forces, and it gives a remarkably good account of the normal liquid from $c.\,0.1\,\mathrm{K}$ down to the superfluid transition. We present here only a very brief outline of the theory. The intention is to describe the physical principles upon which it is built and to indicate briefly how it may be applied in practice. Fuller accounts will be found, for example, in Wilks (1967) and in Baym and Pethick (1978).

The Landau theory employs a perturbation approach to describe an assembly of interacting fermions in the degenerate ($T \ll T_F$) limit where the Fermi sphere is well defined. As in the case of the ideal gas, it is assumed that the wave function of each fermion occupies the entire available volume. Thus, their momenta may be determined in the usual way through the application of periodic boundary conditions (see, for example, Mandl, 1971) and are exactly the same as for the ideal gas. In particular, the momentum at the Fermi surface is still

$$p_F = \hbar k_F = \hbar(3\pi^2 N/V)^{1/3} \qquad (6.8)$$

It is assumed that there will be a one-to-one correspondence between the states of the ideal and interacting systems but that, for the latter, interactions between the atoms will alter the energies of the various states.

Each ^{3}He atom moves subject to the influence of its neighbours. Correspondingly, the movement of any particular atom must cause a moving local distortion in the positions of neighbouring atoms. The atoms of the interacting system do not, therefore, behave like those of the ideal gas and, in particular, they do not behave as free particles of energy $E = \hbar^2 k^2/2m$. To take account of the interactions, the actual atoms are considered to be replaced by an equal number of quasiparticles whose energies E are *defined* by

$$\frac{\delta U}{V} = \int E \, \delta n \, d\tau \qquad (6.9)$$

where δU is the change in the energy of the assembly as a whole, resulting from a small change δn in the distribution function. The integration is taken over all momentum space, so that

$$d\tau = 2 dk_x \, dk_y \, dk_z/(2\pi)^3 \qquad (6.10)$$

where the initial factor of 2 takes account of the two possible spin orientations.

The number of quasiparticles $n(E) \, dE$ which occupy states with energies between E and $E + dE$ is found in the usual way by maximizing the entropy subject to the Pauli exclusion principle and to the requirement that the total number of quasiparticles and the total energy must each be conserved. The resultant distribution function

$$n(E) = \{\exp\left[(E - \mu)/kT\right] + 1\}^{-1} \qquad (6.11)$$

appears at first sight identical to that for the ideal gas. There is a crucial difference, however, in that E in (6.11) depends on the occupation numbers of *all* the permitted states, and hence on n. The chemical potential μ is fixed as usual by the requirement that

$$\frac{N}{V} = \int n \, d\tau. \qquad (6.12)$$

For $T \ll T_F$, the dispersion relation $E(k)$ for quasiparticles near the Fermi surface may to a good approximation be written

$$E = \mu + \left(\frac{\partial E}{\partial k}\right)_F (k - k_F) \qquad (6.13)$$

where, as before, the subscript F refers to quantities evaluated at the Fermi surface. By analogy with the free particle states of an ideal gas, we write

$$\frac{1}{\hbar}\left(\frac{\partial E}{\partial k}\right)_F = \frac{\hbar k_F}{m_3^*} = v_F \qquad (6.14)$$

where m_3^* is now the effective mass and v_F may be identified as the velocity of a quasiparticle at the Fermi surface. Thus,

$$E = \mu + \frac{\hbar^2 k_F}{m_3^*}(k - k_F). \qquad (6.15)$$

The density of states per unit volume at the Fermi surface is given by the same expression as for the ideal gas, but with the quasiparticle effective mass m_3^* incorporated in place of the bare atomic mass, so that

$$D(0) = (m_3^*/\pi^2\hbar^2)(3\pi^2 N/V)^{1/3} \qquad (6.16)$$

To evaluate this expression it is, of course, necessary to determine m_3^* by taking explicit account of the interaction forces.

We now need to consider the effect on any given level of a redistribution of quasiparticles among the other levels. Continuing to ignore, for now, the effect of magnetic fields, we may write

$$E(\mathbf{k}) = E_0(\mathbf{k}) + \int f(\mathbf{k},\mathbf{k}')\,\delta n'\,d\tau' \qquad (6.17)$$

where $E_0(\mathbf{k})$ is the energy of the level with momentum $\hbar\mathbf{k}$ when the assembly is in equilibrium at $T=0$, and $E(\mathbf{k})$ is the energy of the same level when the occupation numbers of the other levels (with momentum $\hbar\mathbf{k}'$) have changed by a small amount $\delta n'(\mathbf{k}')$. Thus, the function

$$f(\mathbf{k},\mathbf{k}') = \frac{\partial^2 U}{\partial n(\mathbf{k})\,\partial n'(\mathbf{k}')} \qquad (6.18)$$

effectively specifies how the energy of the level in question (with momentum $\hbar\mathbf{k}$) has been modified by the change in occupations of all the other levels. In practice, the only values of $\mathbf{k}$ and $\mathbf{k}'$ which are of interest are those close to the Fermi surface, so that $k \simeq k' \simeq k_F$. Hence $f(\mathbf{k},\mathbf{k}')$ must be a function only of the angle Θ between $\mathbf{k}$ and $\mathbf{k}'$, and may therefore be written as $f(\Theta)$. For convenience, Landau introduces a new function $F(\Theta)$, defined in terms of $f(\Theta)$ and the density of states at the Fermi surface (6.16), such that

$$F(\Theta) = D(0)f(\Theta). \qquad (6.19)$$

He then expands F in terms of Legendre polynomials

$$F(\Theta) = \sum_n F_n P_n(\cos\Theta) = F_0 + F_1\cos\Theta + F_2\frac{3\cos^2\Theta - 1}{2} + \cdots \qquad (6.20)$$

where the F_n are constants. To take account of the interaction of the nuclear spins with a magnetic field, he introduces two new functions $g(\Theta)$ and $G(\Theta)$ in a very similar way and, as in the case of F, he expands G as

$$G(\Theta) = \sum_n G_n P_n(\cos \Theta) = G_0 + G_1 \cos \Theta + G_2 \frac{3\cos^2 \Theta - 1}{2} + \cdots \quad (6.21)$$

The coefficients F_n and G_n are unknown *a priori* and are to be determined by experiment. They are referred to as *Landau parameters*. It is found in practice that the first few terms in (6.20) and (6.21) are sufficient to provide an excellent description of the liquid and that, for many purposes, only F_0, F_1 and G_0 need be considered. The experimental values given in Table 6.1 show by the increase of their absolute magnitudes with pressure that, not surprisingly, the effect of the interactions becomes more pronounced as the atoms are pushed correspondingly closer together.

With the interaction function conveniently parameterized in this way, it becomes possible to calculate explicitly almost all of the properties of the interacting Fermi liquid. For a detailed discussion, see Wilks (1967) or Baym and Pethick (1978). In general, it is found that the temperature-dependences of the various quantities are the same as for the ideal Fermi gas but that their absolute magnitudes differ to a considerable extent, in agreement with the experimental observations of the preceding section. We now quote a few examples of the type of result that may be obtained.

The quasiparticle effective mass, in terms of F_1, is

$$m_3^* = m_3(1 + \tfrac{1}{3}F_1). \quad (6.22)$$

The specific heat per unit volume is given by

$$C = \tfrac{1}{3}\pi^2 k_B^2 D(0)T$$

where $D(0)$, the density of states at the Fermi surface, is given by (6.16). Thus C is the same as for the non-interacting system, except that m_3^* is used in (6.16) rather than the bare atomic mass m. Hence

$$C = \frac{m_3^*}{m_3} C^{\text{ideal}} \quad (6.23)$$

where C^{ideal} is the specific heat of the ideal gas, and it may be noted that C

Table 6.1 Landau parameters for liquid ³He (taken from Wheatley, 1975)

	P (bar)		
	0	15	30
F_0	10.07	45.62	82.13
F_1	6.04	11.01	14.58
G_0	−2.69	−2.92	−2.95

depends on F_1 through m_3^*. The spin susceptibility

$$\chi = \frac{m_3^*}{m_3}[1 + \tfrac{1}{4}G_0]^{-1}\chi^{\text{ideal}} \tag{6.24}$$

where χ^{ideal} is the ideal gas value, depends on F_1 and G_0. The negative value found for G_0 (Table 6.1) corresponds to its being easier to align the nuclear spins of liquid ^{3}He than those of an ideal Fermi gas. The velocity of sound u (for temperatures above those of collisionless sound: see below) is given by

$$u^2 = \frac{\hbar^2 k_F^2}{3m_3 m_3^*}(1 + F_0) \tag{6.25}$$

and thus depends on F_0 and F_1. By measurement of these and other properties of liquid ^{3}He it has been possible not only to obtain a set of values for the Landau parameters, but to show that *consistent* values are obtained from very different types of experiment. This may be construed as powerful evidence in support of the general correctness of Landau's approach.

Sound propagation in a Fermi gas is of particular interest at very low temperatures where the average time τ between collisions (see above) becomes long. For any given frequency ω there will be a characteristic temperature at which ω and τ^{-1} become equal. At lower temperatures than this, when

$$\omega\tau \gg 1 \tag{6.26}$$

the propagation of ordinary sound in an ideal Fermi gas clearly becomes

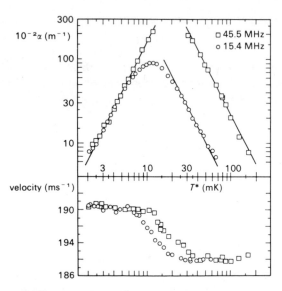

Figure 6.7 The amplitude attenuation coefficient α and the velocity of sound at two different frequencies in pure liquid ^{3}He, each as a function of the magnetic temperature T^*. (After Wheatley, 1970.)

impossible, because the collisions simply do not occur fast enough. For the interacting Fermi liquid, however, a novel 'collisionless' sound propagation mode known as *zero sound* may still be excited. According to Landau's (1957) theory, an increase in the propagation velocity, and a maximum in the attenuation, should be observed as the sample is cooled through the region where ordinary sound gives way to zero sound. This prediction has been elegantly and convincingly verified by experiment (Fig. 6.7).

6.4 Superfluid phases of ³He

Once it had been established that liquid ³He behaves as an interacting Fermi gas it was natural to ask whether, at a sufficiently low temperature, the liquid would undergo a BCS-like pairing transition analogous to that of the electron gas in a superconductor (§4.4). The general consensus prior to the experiments was that such a transition was indeed to be anticipated, but that the result would not necessarily be a superfluid. In the end, the long-awaited transition was detected experimentally by Osheroff *et al.* (1972) using a Pomeranchuk (compressional cooling, §7.3) apparatus at Cornell University. At first, the observed transition was believed to correspond to a phase change in the solid ³He that was also present in the cell, but this initial attribution was soon shown to be erroneous. It was quickly established that the 'new phases' of liquid ³He which they had discovered do, in fact, display superfluidity. Superfluid ³He, being both magnetic and anisotropic, and existing as three quite distinct phases (rather than the single phase which had been anticipated), is a great deal more complicated than superfluid ⁴He, which is isotropic and magnetically inert. Nonetheless, the superfluid phases of the two isotopes have much in common with each other, as well as many interesting and important differences. In the necessarily brief and somewhat qualitative account of superfluid ³He which follows we will make frequent reference to analogous properties of HeII described in chapter 5. For further information, reference may be made to the excellent reviews by Leggett (1975), Wheatley (1975, 1978), Brinkman and Cross (1978), Wölfle (1978) and Lee and Richardson (1978), the last article being particularly accessible to the non-specialist. Superfluid ³He is at present under intense theoretical and experimental investigation, and the reader should bear in mind that the picture built up thus far will certainly be substantially extended, and probably to some extent modified, over the next few years.

The low-temperature part of the ³He phase diagram (Fig. 1.9*b*) is profoundly affected by a magnetic field. In zero magnetic field, there are two distinct superfluid phases, ³He-A and ³He-B. As indicated in Fig. 6.8*a* there is a unique point, known as the polycritical point (PCP) at which the A and B phases and the normal liquid are all in mutual equilibrium. When a magnetic field is applied, however, the situation changes dramatically (Fig. 6.8*b*). The PCP then no longer exists and, instead, a region of A-phase extends right down

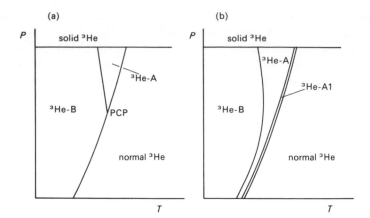

Figure 6.8 Schematic diagram illustrating the effect of a magnetic field on the low-temperature region of the ^{3}He phase diagram. In (a) there is no magnetic field, and in (b) a finite magnetic field has been applied.

to zero pressure. At the same time, a completely new superfluid phase (designed the A1-phase) appears, separating the A-phase from the normal liquid. As the field is further increased, the A-phase region, and to a lesser extent, that of the A1 phase, grow wider. Simultaneously the B-phase retreats towards lower temperatures until, in fields above $c.\,0.6\,T$, it has been replaced entirely with A-phase.

The normal → A transition is found to be a second-order phase transformation, characterized by a *finite* discontinuity in the specific heat (Fig. 6.1*b*). It clearly bears a close resemblance to $C(T)$ for the normal → superconducting transition for the electron gas in a metal (Fig. 4.1*a*), but is entirely different from the λ-transition in liquid ^{4}He (Fig. 5.1). On application of a magnetic field, the normal → A transition splits into a pair of second-order transitions (normal → A1, A1 → A) which occur at slightly different temperatures. The A → B transition, on the other hand, is entirely different in character, for it is a first-order phase transformation (like boiling or freezing) with an associated latent heat and, unlike the normal → A (or A1 → A) transitions, a considerable measure of supercooling can occur before the transition takes place.

The interaction responsible for the pairing transition is a complicated one, made up from a number of different contributions. These include the van der Waals' force, an important indirect force due to what are known as *spin fluctuations,* and an exchange force (the direct dipole–dipole interaction between the tiny nuclear magnetic moments being negligible by comparison). The net result is a Lennard–Jones-like potential which is strongly repulsive at short range but which can be attractive at larger separations. Unlike the electron gas in a superconductor, where the Cooper pairs (§4.4) consist of fermions with opposed spins ($S = 0$) and zero mutual orbital angular momentum (S-wave pairing, with $L = 0$), pairing in superfluid ^{3}He occurs between atoms with equal spin ($S = 1$) and finite orbital angular momentum. Equal spin pairing is in fact to be expected in view of the nearly ferromagnetic character of the normal liquid, as revealed by its enhanced spin susceptibility over that of an ideal Fermi gas (see §6.3), which favours the formation of small clusters of atoms with parallel spins. Because the two atoms of the pair are then in identical spin states, the Pauli principle requires that L be odd in order that the overall wavefunction should remain antisymmetric. The measured properties of the superfluid phases indicate that $L = 1$, that is, *P-wave pairing*. The two atoms of a pair can therefore be envisaged as orbiting around each other: centrifugal forces tend to keep them apart, so that their binding is not jeopardized by the hard-core repulsion that occurs when they get too close together.

For Cooper pairs with $S = 1, L = 1$, there is clearly a large number of possible states in which the liquid as a whole can exist even if, for now, we ignore the possible influences of walls, electric and magnetic fields, temperature gradients, flow fields and so on. For $S = 1$, the possible spin states are $S_z = 1, 0$, or -1, corresponding to spin orientations ↑↑, ↓↑ or ↓↓ (where ↑↓ should be taken to indicate the appropriate normalized linear sum over ↑↓ and ↓↑ states). The experiments seem to show that all three spin species are present in the B-phase superfluid which corresponds to the (BW) state originally proposed by Balian and Werthamer (1963). The A-phase, on the other hand, apparently includes only ↑↑ and ↓↓ pairs, corresponding to the (ABM) state first suggested by Anderson and Morel (1961) and Anderson and Brinkman (1973). In the A1 phase, it appears that only one of the latter species is present; but it is currently unknown whether it is the ↑↑ or ↓↓ pairs. The relationships between spin and orbital angular momenta adjust themselves in each case so as to minimize the free energy of the liquid as a whole, and turn out to be entirely different for the A and B-phases.

In the A-phase (and the A1-phase) the orbital angular momentum vector for every pair is orientated in the same direction, so that the liquid as a whole may be presumed to carry a macroscopic angular momentum **l** which is non-zero (but whose magnitude remains a matter for debate). Liquid ^{3}He-A has consequently been described as an *orbital ferromagnet*. The character of ^{3}He-B is strikingly different, in that the pair orbital momenta are orientated in different directions for different points on the Fermi surface, so that the net angular momentum of the B-phase in its ground state is zero.

In common with other macroscopic quantum systems, superfluid ^{3}He may be characterized by an order parameter (§1.6) which is effectively the macroscopic analogue of a single-particle wavefunction. In view of the intrinsically anisotropic character of the triplet-state pairs, it is not at all surprising that the order parameter for superfluid ^{3}He has turned out to be a great deal more complicated than the scalar quantities that may be used to describe HeII or the BCS electron gas in a superconductor. It may be written in the form of a vector **d(k)**, whose physical significance is that $|\mathbf{d(k)}|$

represents the pair condensate amplitude at the point where the direction **k** intersects the Fermi surface. The direction and magnitude of **d** are quite different for ^{3}He-A and ^{3}He-B.

The Fermi surfaces and energy gaps, and the equilibrium relationships between some of the vectors characterizing the two phases, are shown in Fig. 6.9. In the A-phase, **d(k)** lies in the *same* direction for all **k**. The energy of the liquid is minimized if this direction lies parallel to **l** and if both vectors are orientated perpendicular to an applied magnetic field **B**. The net magnetization **M** of the liquid lies parallel to **B**. Nodes in **d** occur as shown at two points on the Fermi surface, where the pair condensate amplitude and the energy gap $\Delta(T)$ fall correspondingly to zero.

In the B-phase, $|\mathbf{d(k)}|$ is independent of **k**, so that the energy gap $\Delta(T)$ is isotropic, but the direction of **d(k)** varies over the Fermi surface. The structure of **d(k)** may be envisaged as follows. First the **d** vectors should be imagined to stick out radially, all over the Fermi surface. Secondly, in order to minimize the magnetic dipole energy, all the **d** vectors are rotated through the 'magic angle' $\cos^{-1}(-\tfrac{1}{4}) \simeq 104°$, about an arbitrary axis **N**. This procedure corresponds, in effect, to an adjustment of the orbital axis of each pair relative to its spin axis. It will have maximum effect on the **d** vectors in the plane perpendicular to **N**, which are moved into an almost tangential orientation, as shown (Fig. 6.9), and will have no effect on those on the **N** axis. Finally, again in the interests of

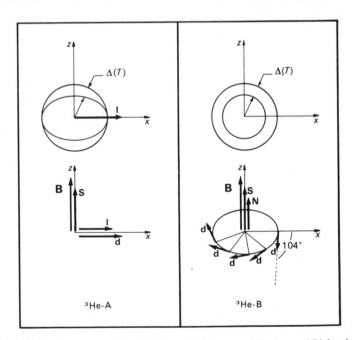

Figure 6.9 Schematic representation of ^{3}He superfluid vectors (after Lee and Richardson, 1978). For explanation, see text.

dipolar energy minimization, **N** orientates itself parallel to an applied magnetic field **B** and to the consequent magnetization **M**. The rotation axis **N** is closely analogous to the directorix vector of a liquid crystal.

The onset of superfluidity in liquid ^{3}He as the temperature is being reduced is particularly striking because of the high viscosity of the normal liquid just above T_c (cf. Fig. 6.2), at which point it has become comparable with light machine oil. The remarkable change in behaviour that occurs at T_c has been dramatically demonstrated by measurements of the damping force experienced by an immersed wire vibrating in the liquid (Fig. 6.10). The width of the resonance peak, which provides a measure of the damping, was observed to decrease by more than five orders of magnitude as T was reduced from T_c to the minimum temperature of c. 140 μK. The fact that the damping fell to a value which was *less* than that measured in a good vacuum at room temperature effectively allays any lingering doubts about the genuineness of the superfluidity of liquid ^{3}He.

Many of the superfluid properties of HeII (chapter 5) have now also been demonstrated in superfluid ^{3}He. In particular, it has been shown that the liquid may be described in terms of a two-fluid hydrodynamics comparable with that developed for HeII (§5.2–5.5). The ^{3}He superfluid component is thus to be associated with the Cooper pair condensate and the normal fluid component with unpaired fermions in thermally excited states. There are added complications arising from the magnetic properties of the liquid and from 'textures' (see below). The two-fluid hydrodynamics of the A-phase is

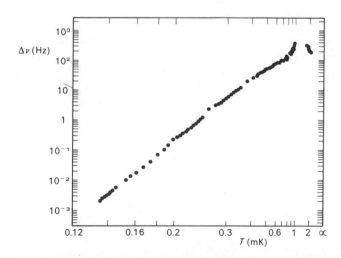

Figure 6.10   The damping experienced by a vibrating wire immersed in liquid ^{3}He under a pressure of 0.0 bar. The measured frequency width $\Delta\nu$ of the resonance, plotted as a function of temperature T, implies a reduction in the damping force of more than five orders of magnitude when the sample is cooled from T_c (1.04 mK) to the minimum temperature reached (0.14 mK). (After Guénault *et al.*, 1983.)

particularly complicated because of its anisotropy, one aspect of which is that quantities such as the superfluid density ρ_s vary with direction in the liquid. The temperature dependence of ρ_s has been measured (Berthold *et al.*, 1976) by means of an experiment analogous to that of Andronikashvili for HeII (cf. Fig. 5.7), with the results shown in Fig. 6.11. It is clear that the measured values of ρ_s/ρ are strongly influenced by the orientation of the applied magnetic field **B**. Such behaviour is, of course, only to be expected if **l** and **d** adjust themselves so as to lie perpendicular to **B**, as sketched in Fig. 6.9(*a*).

In fact, the exact orientations of the vectors are influenced by a whole range of factors, and they are not usually able to adopt the idealized arrangement of Fig. 6.9 throughout the whole body of the liquid. Cooper pairs adjacent to surfaces are constrained to orbit in a plane parallel to the surface, in order to avoid pair-breaking collisions with the surface itself. Thus, as far as possible, **l** in ^{3}He-A tends to meet the container walls at right angles. In cases where the influences of the walls and of **B** are in competition, **l** will be perpendicular to the walls at very close distances; but **B** will tend to dominate in the interior of the liquid, with **l** turning perpendicular to **B** as in Fig. 6.9. The liquid consequently acquires a *texture*. The characteristic length for a change in orientation is of order $10\,\mu$m for ^{3}He-A so that, except in containers of very small dimension, **B** is dominant throughout most of the liquid. In ^{3}He-B, it is the **N** vector which tends to lie perpendicular to surfaces and parallel to **B**. The characteristic bending length for **N** is very much larger than that for **l** in ^{3}He-A and, depending on the magnitude of **B**, may be as large as several mm. For this reason, textural effects in ^{3}He-B are somewhat easier to study than those in ^{3}He-A.

Other factors which are expected to influence the textures include heat currents, which induce normal fluid/superfluid counterflow much as in HeII, and electric fields, although the effect of the latter appears to be unexpectedly weak. Even in the absence of currents or externally applied fields it is, of course, topologically impossible for **l** (or **N**) to meet *all* walls perpendicularly in a real container without singularities occurring somewhere within the texture. Two possible configurations that have been suggested for ^{3}He-A within a spherical container are sketched in Fig. 6.12.

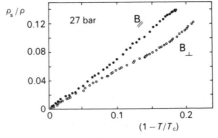

Figure 6.11   The relative superfluid density of ^{3}He-A for two different orientations of the applied magnetic field as a function of reduced temperature. (After Berthold *et al.*, 1976.)

Figure 6.12 Two examples of possible textures taken up by ³He-A inside a sphere. The total energy is minimized if the l vectors (illustrated) meet the walls perpendicularly; but it is topologically impossible for this requirement to be satisfied everywhere and one or more singularities are therefore inevitable.

Just as in the cases of HeII and the electron gas in superconductors, it is to be expected that the superfluidity of liquid ³He will break down above a characteristic critical velocity. The Landau arguments of §5.4 may readily be applied to ³He, yielding a pair-breaking critical velocity of $v_L = \Delta/\hbar k_F$. Taking for the energy gap Δ the BCS value at $T = 0$ (§4.4) of $1.76\,k_B T_c$ we find that v_L should lie within the range $c.\ 30 < v_L < 70\,\mathrm{mm\,s^{-1}}$ depending on pressure; which is to be compared with $45 < v_L < 60\,\mathrm{ms^{-1}}$ in HeII. Ion experiments in superfluid ³He have not yet been carried to low enough temperatures to allow a rigorous test of this prediction, but the data of Ahonen et al. (1976) appear to suggest that a characteristic very similar to that of Fig. 5.16 will be obtained, with v_L falling close to the expected value. A number of flow experiments have also been performed, usually yielding one or more distinct critical velocities of order mm or $\mathrm{cm\,s^{-1}}$.

Notwithstanding the many features which superfluid ³He shares in common with HeII and superconductors, its magnetic properties make it unique. These properties were of crucial importance in aiding the identification of the A and B-phases respectively with the ABM and BW states described above. The static magnetic susceptibility $\chi(T)$ of ³He-A is equal to that χ_F of the normal liquid above T_c (cf. Fig. 6.4) and is temperature-independent. For ³He-B, $\chi(T)$ falls below χ_F, which must of course be expected in view of the formation of a proportion of nonmagnetic ↑↓ pairs.

Both ³He-A and ³He-B possess *characteristic frequencies* $\Omega_A(T)$ and $\Omega_B(T)$ corresponding to oscillations of the **d** vector about its equilibrium orientation (Fig. 6.9). These frequencies may be measured by nuclear magnetic resonance techniques. For the A-phase, the frequency $\omega(T)$ of the conventional transverse resonance is found to be shifted away from the Larmor frequency γB according to the relation

$$[\omega(T)]^2 = (\gamma B)^2 + [\Omega_A(T)]^2 \tag{6.27}$$

where Ω_A starts from zero at T_c and increases with decreasing temperature as shown in Fig. 6.13. The A-phase also exhibits an extraordinary longitudinal NMR resonance (that is, with the static and radiofrequency magnetic fields parallel to each other rather than in the usual perpendicular orientation) which yields $\Omega_A(T)$ directly. Measurement of $\Omega_B(T)$ is harder because, in the B-

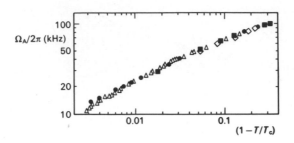

Figure 6.13 The characteristic frequency of ^{3}He-A near the melting pressure as a function of reduced temperature, from a number of different types of experiment: longitudinal resonance (filled circles and squares) by Gully *et al.* (1976) and Osheroff and Brinkman (1974) respectively; transverse resonance (diamonds) by Osheroff and Brinkman (1974); magnetic parallel ringing (triangles) by Webb *et al.* (1974). (After Gully *et al.*, 1976.)

phase, there is usually no frequency shift away from γB; but it can still be deduced by means of NMR through the use of special techniques which force **N** away from the direction of **B**.

The characteristic frequencies of both phases may also be measured (Fig. 6.13) by exploitation of the remarkable phenomenon of parallel ringing. The applied magnetic field is stepped up or down by a small increment, with the result that the net magnetization of the ^{3}He sample starts oscillating at Ω_A or Ω_B. The oscillation, which is only weakly damped, continues until the relevant vectors have settled back into their equilibrium configuration (Fig. 6.9) once more. An equivalent description of the phenomenon can be made in terms of the transfer, by tunnelling, of Cooper pairs between the two separate but weakly coupled and interpenetrating superfluids comprised of the ↑↑ and ↓↓ pairs. This process amounts to a kind of internal Josephson effect (cf. §4.6). Its existence was predicted by Leggett (1973), and sought and found subsequently by Osheroff and Brinkman (1974); it has therefore been called the *Leggett effect*. Ringing has not been observed in the A1 phase where only one species of pairs is present and the process cannot, therefore, take place.

In view of the complicated nature of superfluid ^{3}He, it is clear that an extensive range of collective oscillatory modes may be anticipated. Zero, first, second and fourth sound have all now been detected. Zero sound is found to exhibit a large excess attenuation close to T_c (Lawson *et al.*, 1974) which may in part be attributed to pair-breaking by zero sound quanta when $\Delta(T)$ is small. Second sound at these low temperatures has an exceedingly small velocity (5.41) and it consequently suffers a large attenuation α_2 (since $\alpha_2 \propto u_2^{-3}$), which makes it extremely difficult to observe. It has nonetheless been detected and studied in the A1 phase (Corruccini and Osheroff, 1980) where it takes on some of the character of a spin wave. Fourth sound (Kojima *et al.*, 1974) has been used to determine ρ_s/ρ and was also of particular importance in providing the first unambiguous demonstration of the superfluidity of the new phases of ^{3}He.

Numerous other modes (Wölfe, 1977), including orbital waves and spin waves, are also possible.

6.5 Liquid ^{3}He–^{4}He solutions

Dilute solutions of ^{3}He in HeII at low temperatures have properties that are, in many respects, much closer to those of the normal phase of pure liquid ^{3}He than to those of pure HeII. They are of particular importance for two main reasons. First, a dilute ^{3}He–^{4}He solution constitutes an almost ideal Fermi gas (see below) whose Fermi temperature can be adjusted at will simply by varying the ^{3}He concentration. Consequently, they are of considerable interest in their own right and have made it possible to verify in quantitative detail many of the predictions of quantum statistical mechanics. Secondly, dilute solutions are of crucial importance to millikelvin technology. Their special properties provide the basis of operation of the helium dilution refrigerator (§7.3), which now constitutes the workhorse for virtually all experimental physics below 0.3 K.

The phase-separation diagram for helium is shown in Fig. 6.14. Above 0.86 K the two isotopes are miscible in all proportions; but the liquid will not be a superfluid if it contains more than a certain proportion of ^{3}He. When mixtures containing more than 6% of ^{3}He are cooled to temperatures below 0.86 K, they eventually separate into two components: a ^{3}He-rich phase floating on top of a denser ^{4}He-rich phase, separated by a visible interface. For any given temperature, the equilibrium concentrations of the two components are given by the points at which the corresponding horizontal line, for example AA', cuts the phase-separation curve. The vicinities of both the lambda line and of the *tricritical point*, where the lambda line and the phase-separation curve meet, have been the subject of detailed investigation (Ahlers, 1976). At very low temperatures the solubility of ^{4}He in ^{3}He tends to zero, whereas that

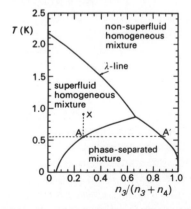

Figure 6.14 The phase diagram for liquid ^{3}He–^{4}He mixtures under their saturated vapour pressure where n_3 and n_4 are the number densities of ^{3}He and ^{4}He respectively and T is the temperature. The finite solubility of ^{3}He in ^{4}He at $T = 0$ should be noted.

of ^{3}He in ^{4}He remains finite, approaching a limiting value of about 6% under the saturated vapour pressure.

This non-zero solubility at $T = 0$ implies that a ^{3}He atom must have a lower energy when placed in pure liquid ^{4}He than it would have in pure liquid ^{3}He. Being subject to Fermi statistics, ^{3}He atoms added sequentially to HeII at $T = 0$ must go into successively higher energy states. Eventually, therefore, a concentration will be reached such that there is no energy advantage for a ^{3}He atom by being in liquid ^{4}He rather than in liquid ^{3}He, and it is this limit which corresponds to the maximum solubility observed experimentally.

The properties of dilute ($\leqslant 6\%$) ^{3}He–^{4}He solutions may conveniently be considered in three regimes of temperature. First, above T_λ (noting from Fig. 6.14 that T_λ will be slightly depressed compared to its value in pure ^{4}He), the properties of a solution approximate to those of pure HeI. This is, of course, only to be expected given that the pure liquids ^{3}He and ^{4}He behave so similarly in this temperature range. Secondly, for $0.5 \lesssim T \lesssim T_\lambda$, the solution possesses most of the properties of HeII but modified, in some cases very substantially, by the ^{3}He. The liquid is well described by a suitably modified version of the two-fluid model (§5.2) in which explicit account is taken of the additional, temperature-independent, contribution made by the ^{3}He to the normal fluid component. Appropriate additions can be made to the equations of motion (5.17–5.22) as discussed, for example, by Khalatnikov (1965). For most properties, the influence of the ^{3}He is minimal just below T_λ, but increases rapidly as T is reduced and the relative size of the ^{3}He contribution to the total entropy and normal fluid density grows, eventually dominating that due to the phonons and rotons.

The effective thermal conductivity of a solution is markedly reduced below that of pure HeII. This can be understood qualitatively in terms of the thermal counterflow sketched in Fig. 5.8. The ^{3}He moves with the normal fluid component, of which it is a part. It therefore tends to accumulate at the colder end of the apparatus, where the rest of the normal fluid component becomes converted to superfluid (corresponding to the absorption of phonons and rotons at the cold sink) and flows back towards the heater. Careful account must, of course, be taken of this *heat flush* of ^{3}He atoms in all experiments on superfluid ^{3}He–^{4}He solutions. In practice, a ^{3}He concentration gradient is created as a result of the scattering of back-diffusing ^{3}He atoms by the oncoming excitations which constitute the normal fluid component. This scattering reduces the efficacy of the thermal conduction process and results in a corresponding temperature gradient. Heat flush has been exploited as the basis of a technique for the complete removal of ^{3}He from helium of the natural isotopic ratio (see §8.5).

Thirdly, as T falls below c. 0.5 K, a new regime is entered where the ^{3}He is best considered as an ideal gas and the presence of the superfluid ^{4}He can, for many purposes, be ignored. At these low temperatures, the entropy contribution from a few percent of ^{3}He vastly outweighs that of the residual phonons and rotons. The mean free path for the scattering of ³He atoms by phonons or rotons has correspondingly become very long. The liquid ⁴He just constitutes a superfluid background or 'mechanical vacuum' within which the ³He atoms can move quite without let or hindrance. Virtually the only effects arising from the presence of the ⁴He are: to endow the ³He atoms with a hydrodynamic effective mass m_3^* which is about 2.5 times larger than the bare atomic mass; and, of course, to prevent the ³He atoms condensing to form a liquid, as they do in pure ³He. Otherwise, the container may be imagined as filled only with the dilute gas of ³He atoms. The dispersion relation for the ³He atoms in a dilute solution, relating their energy E and momentum $\hbar k$, is

$$E = E_0 + \hbar^2 k^2 / 2m_3^* \tag{6.28}$$

where E_0 is the potential energy of one ³He atom in pure liquid ⁴He. Equation (6.28) is thus the analogue of (5.44) and (5.45) for phonons and rotons respectively.

The discussion above of the ideal Fermi gas (§6.1) is applicable *a fortiori* to dilute ³He–⁴He solutions. As an example, we consider the behaviour of the specific heat. At 0.5 K, where $T \gg T_F$, Maxwell–Boltzmann statistics are applicable and the molar specific heat is found to lie very close to the (temperature-independent) classical gas value of $\frac{3}{2}R$. At lower temperatures, where $T \ll T_F$ and the influence of Fermi statistics is expected to be paramount, the measured limiting behaviour of $C(T)$ is linear in T, as shown in Fig. 6.15, consistent with the predictions of (6.1). In sharp contrast to the case of pure liquid ³He, the values of m_3^* that may be deduced by fitting the ideal Fermi gas model to a variety of different types of experimental data are found to be consistent with each other.

An attractive interaction, albeit an exceedingly weak one, does in fact exist between the ³He quasiparticles, owing to the greater volume which they occupy in the liquid (on account of their larger zero-point energy) as compared to the ⁴He atoms. It is possible that, at a sufficiently low temperature, this interaction may give rise to a BCS-like pairing transition in states of non-zero orbital angular momentum, much as in pure liquid ³He. Such a 'superfluid

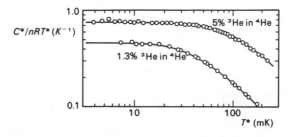

Figure 6.15 Ratio of the molar heat capacity to RT^* for two dilute ³He–⁴He solutions, where T^* is the magnetic temperature and R is the gas constant (after Wheatley, 1970).

within a superfluid' would, of course, be a system of very considerable interest. The transition has been carefully sought, so far down to about 0.3 mK, but without success.

6.6 Other fermion fluids

Thus far, we have considered the low-temperature behaviour of three different types of fermion fluid: the electron gas in metals, liquid ^{3}He, and liquid ^{3}He–^{4}He solutions. In each case, the properties of the system are profoundly influenced by Fermi–Dirac statistics and, for the first two (though not, of course, for every metal), a pairing transition to a superfluid state occurs at low enough temperatures. It is of interest to consider what other systems in nature might be expected to display comparable behaviour.

One possible candidate is monatomic deuterium (D), which must form a fermion fluid (cf. §5.8 on the boson fluid monatomic hydrogen). At high enough densities and/or low enough temperatures, such that $T \ll T_F$, one may hope to observe characteristic Fermi behaviour and even, perhaps, to detect a pairing transition if it occurs. It appears, however, that D is relatively hard to stabilize, so that the highest densities currently being achieved are a hundred times smaller than for H. Clearly, a number of problems remain to be overcome. The present state of this interesting research field is reviewed by Silvera (1982).

Monolayers of ^{3}He on atomically flat substrates constitute 2-D fermion gases or liquids. Their low-temperature properties differ considerably from those of ^{4}He monolayers, probably because of the different quantum statistics that are applicable. Details will be found in the review by Dash and Schick (1978).

White dwarf stars are believed to represent the end product of stellar evolution for stars of small to medium mass. They consist mainly of ionized helium, typically at $T \sim 10^7$ K, and thus comprise alpha particles and electrons. Because of their large density ($c.\ 10^{10}$ kg m^{-3}) the Fermi temperature is typically 10^{11} K. Thus $T \ll T_F$ and, notwithstanding the large absolute value of T, the electron gas must be treated as a fermion fluid in the extreme low-temperature limit. By considering the star in this way, taking due account of the fact that the electrons are moving at relativistic velocities, one can readily show that its equilibrium radius *decreases* with increasing mass, and that there is a critical mass ($c.\ 1.4$ solar masses) above which the radius becomes imaginary and a stable configuration as a white dwarf star does not exist. This result is in accord with astronomical observations: for details see, for example, Huang (1963).

For stars of more than 1.4 solar masses, where the pressure due to the relativistic electron gas is insufficient to sustain them against gravitational collapse, a new stable configuration may, however, still be attained, provided that the star is not too massive, as a neutron star. Pulsars, of which a large number have now been catalogued, are believed to consist of rapidly rotating

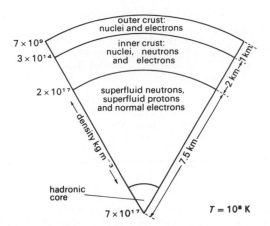

Figure 6.16 Calculated structure for a medium-weight neutron star (after Pines, 1971).

neutron stars, with periods ranging from a few ms to several seconds. A possible structure (Pines, 1971), is sketched in Fig. 6.16. The star consists mainly of neutrons, has an average density of $c.$ 10^{17} kgm^{-3} and a temperature of $c.$ 10^8 K. The neutrons are of course fermions and, for the density in question, the Fermi temperature will be $c.$ 10^{11} K, so that $T \ll T_F$. Like the electrons in the white dwarf star, therefore, the neutrons may be treated as a fermion liquid in the extreme low-temperature limit. It is believed that they will exist in a paired BCS-like superfluid state which, depending on the local density, may be either S-wave (like electrons in a superconductor, §4.4), or P-wave (like superfluid ³He, §6.4). Just as in HeII, it is expected that the neutron superfluids will be unable to rotate like solid bodies. Most of the angular momentum of a neutron star is believed to reside in an array of quantized vortex lines, closely analogous to those of HeII (§5.5).

Bibliography

References

Ahonen, A.I., Kokko, J., Lounasmaa, O.V., Paalanen, M.A., Richardson, R.C., Schoepe, W., and Takano, Y., *Phys. Rev Lett.* **37**, 511 (1976).
Anderson, P.W. and Morel, P. *Phys. Rev.* **123**, 1911 (1961).
Anderson, P.W. and Brinkman, W.F. *Phys. Rev. Lett.* **30**, 1108 (1973).
Balian, R. and Werthamer, N.R. *Phys. Rev.* **131**, 1553 (1963).
Berthold, J.E., Giannetta, R.W., Smith, E.N. and Reppy, J.D. *Phys. Rev. Lett.* **37**, 1138 (1976).
Betts, D.S. *Refrigeration and Thermometry Below One Kelvin*. Sussex University Press (London, 1976).
Corruccini, L.R. and Osheroff, D.D. *Phys. Rev. Lett.* **45**, 2029 (1980).
Guénault, A.M., Keith, V., Kennedy, C.J., Miller, I.E. and Pickett, G.R. *Nature* **302**, 695 (1983).
Gully, W.J., Gould, C.M., Richardson, R.C. and Lee, D.M. *J. Low Temp. Phys.* **24**, 563 (1976).
Halperin, W.P., Archie, C.N., Rasmussen, F.B., Alvesalo, T.A. and Richardson, R.C. *Phys. Rev.* **B13**, 2124 (1976).

Huang, K. *Statistical Mechanics*. Wiley, New York (1963).
Khalatnikov, I.M. *An Introduction to the Theory of Superfluidity*. Benjamin, New York (1965).
Kojima, H., Paulson, D.N. and Wheatley, J.C. *Phys. Rev. Lett.* **32**, 141 (1974).
Landau, L.D. *Zh. éksp. teor. Fiz.* **30**, 1058 (1956); translated in *Soviet Phys. J.E.T.P.* **3**, 920 (1957).
Landau, L.D. *Zh. éksp. teor. Fiz.* **32**, 59 (1957); translated in *Soviet Phys. J.E.T.P.* **5**, 101 (1957).
Lawson, D.T., Gully, W.J., Goldstein, S., Richardson, R.C. and Lee, D.M. *J. Low Temp. Phys.* **15**, 169 (1974).
Leggett, A.J. *J. Phys. C: Solid St. Phys.* **6**, 3187 (1973); and *Phys. Rev. Lett.* **31**, 352 (1973).
Osheroff, D.D., Richardson, R.C. and Lee, D.M. *Phys. Rev. Lett.* **28**, 885 (1972); Osheroff, D.D., Gully, W.J., Richardson, R.C. and Lee, D.M. *Phys. Rev Lett.* **29**, 920 (1972).
Osheroff, D.D. and Brinkman, W.F. *Phys. Rev. Lett.* **32**, 584 (1974).
Pines, D. *Proc. 12th Int. Conf. on Low Temp. Phys.* (at Kyoto, 1970), p7; Keigaku Publishing Co., Tokyo (1971).
Thomson, A.L., Meyer, H. and Adams, E.D. *Phys. Rev.* **128**, 509 (1962).
Webb, R.A., Kleinberg, R.L. and Wheatley, J.C. *Phys. Rev. Lett.* **33**, 145 (1974).

Further reading

Ahlers, G. 'Experiments near the superfluid transition in ^{4}He and in ^{3}He–^{4}He mixtures', p. 85ff. in Bennemann and Ketterson (1976), q.v.
Baym, G. and Pethick, C. 'Landau Fermi—liquid theory and low temperature properties of normal liquid ^{3}He', p. 1ff; and 'Low temperature properties of dilute solutions of ^{3}He in superfluid ^{4}He', p. 123ff; both in Bennemann and Ketterson (1978), q.v.
Bennemann, K.H. and Ketterson, J.B. (eds.) *The Physics of Liquid and Solid Helium*. Wiley, New York, part I (1976) and part II (1978).
Brewer, D.F. (ed.) *Proceedings of the 1976 Sussex Symposium on Superfluid ^{3}He. Physica 90* B and C (1977).
Brewer, D.F. (ed.) *Progress in Low Temperature Physics*. Vol. VIIA, North-Holland, Amsterdam (1978).
Brinkman, W.F. and Cross, M.C. 'Spin and orbital dynamics of superfluid ^{3}He', p. 105ff. in Brewer (1978), q.v.
Dash, J.G. and Schick, M. 'Helium monolayers', p. 497ff in Bennemann and Ketterson (1978), q.v.
Dobbs, E.R. 'Superfluid Helium Three', *Contemp. Phys.* **24**, 389 ff. (1983).
Lee, D.M. and Richardson, R.C. 'Superfluid ^{3}He', p. 287ff in Bennemann and Ketterson (1978), q.v.
Leggett, A.J. 'A theoretical description of the new phases of liquid ^{3}He'. *Rev. Mod. Phys.* **47**, 331 (1975).
Mandl, F. *Statistical Physics*. Wiley, London (1971).
Silvera, I.F. 'Spin-polarised hydrogen and deuterium: quantum gases' in Clark W.G. (ed.), *Proc. 16th Int. Conf. on Low Temp. Phys.* (Los Angeles, 1981), *Physica 109* and *110B* and *C*, 1499 (1982).
Wheatley, J.C. 'Experimental properties of pure ^{3}He and dilute solutions of ^{3}He in superfluid ^{4}He at very low temperatures', p. 77ff. in Gorter, C.J. (ed.), *Progress in Low Temperature Physics*, vol. VI, North-Holland, Amsterdam (1970).
Wheatley, J.C. 'Experimental properties of superfluid ^{3}He', *Rev. Mod. Phys.* **47**, 415 (1975).
Wheatley, J.C. 'Further experimental properties of superfluid ^{3}He', p. 1ff. in Brewer (1978), q.v.
Wilks, J. *The Properties of Liquid and Solid Helium*. Clarendon Press, Oxford (1967); and an abridged and updated version, *An Introduction to Liquid Helium*, Clarendon Press (1970).
Wölfle, P. 'Collisionless collective modes in superfluid ^{3}He' p. 96ff. in Brewer (1977), q.v.
Wölfle, P., 'Sound propagation and kinetic coefficients in superfluid ^{3}He', p. 191ff. in Brewer (1978), q.v.
Zemansky, M.W. and Dittman, R.H. *Heat and Thermodynamics*. 6th edn., McGraw-Hill, New York (1981).

7 Experimental methods at low temperatures

7.1 Principles of cryostat design

Our objectives in this chapter are two-fold. First, we shall describe the general principles involved in the design and use of cryostats for making measurements on the properties of matter at low temperatures. We shall assume that, as is invariably the case in modern low-temperature laboratories, liquid nitrogen and liquid helium are freely available either by on-site liquefaction or through external purchase. We shall then discuss the physical principles and particular problems associated with various cooling methods that are useful over different temperature ranges.

Techniques for reaching temperatures above 1 K are very well established; most depend in some way on the direct use of ^{4}He in either its liquid or its gaseous state. We shall summarize these methods in §7.2 since most of the applications of low-temperature properties occur in the regime above 1 K. The real frontier of the subject is now, however, at temperatures far below 1 K. There are many laboratories, including those at Lancaster University, where temperatures below 1 mK are routinely obtained. The methods by means of which such temperatures can be achieved depend on physical properties quite different from those of simple evaporation. The singular properties of ^{3}He play a central role, as described below in §7.3, but the very lowest temperatures are reached by adiabatic nuclear demagnetization (§7.4).

The first stage in the commissioning of a new cryostat consists of a careful and detailed evaluation of the functions that the system will be required to perform, including a realistic appraisal of the margins of error that will be acceptable. Overdesign in cryogenics can turn out to be very wasteful of both time and money. With the exception of special-purpose facilities, such as those for high magnetic fields or for millidegree temperatures, it is rare for a single cryostat to be satisfactory for more than one or two different applications. The first questions that must be asked relate to temperature. Over what range will measurements be made? To what accuracy will the temperature need to be maintained constant at values within that range, and for what length of time?

Answers to these questions determine the basic method by which cooling power will be obtained. Experiments at temperatures above about 1 K can be carried out easily in cryostats using liquid ^{4}He. As we have seen, liquid ^{4}He boils under atmospheric pressure at 4.2 K, which is a convenient temperature for investigations that merely require a low temperature environment in order, for example, to minimize thermal effects, such as scattering. Temperatures between 4.2 K and about 1 K can be reached by pumping on liquid ^{4}He in order to reduce the vapour pressure. A common method of maintaining a constant temperature at some point above 4.2 K is to balance the cooling power of the liquid ^{4}He with a deliberate heat leak which is stabilized by a feedback loop. At higher temperatures still, other cooling methods become advantageous. Continuous-flow cryostats through which cooled helium gas is continuously circulated, and miniature internal-work refrigerators designed for specific functions, have gained acceptance in recent years.

For temperatures in the range 1 K down to 0.3 K, a pumped ^{3}He evaporation cryostat may be used. Although exactly the same in physical principle as the pumped ^{4}He cryostat, the ^{3}He system differs considerably in detail because of the very high cost of the material, approximately £100 per litre of gas at STP in 1983, making conservation a crucial element in design. In recent years, the use of a ^{3}He–^{4}He dilution refrigerator has become almost universal for all studies below 0.3 K. Complete systems for a variety of applications can be purchased from at least two different manufacturers, and it is only at the bottom end of the range below 10 mK that there is much divergence in capability and reliability. Dilution refrigerators will be discussed in more detail in §7.3. The problems associated with the definition, measurement and control of temperature are fundamental to low-temperature experiments and will be considered in §7.5.

Once the type of cryostat has been settled, the next important question that must be answered relates to the nature and extent of the connections that are necessary between room temperature and the sample at low temperatures, which will clearly depend on the type of measurement to be made. There are two separate problems. The currents or voltages used in the process of making the measurements may themselves heat up sensitive parts of the experiment, for instance, in measuring the electrical resistance of a semiconducting thermometer. For this reason, low-temperature measurements are always made using the lowest possible sensing currents and voltages, and this requirement often necessitates the use of signal recovery techniques such as signal averaging, boxcar integration, or phase-sensitive detection. It is not only electrical radiation that must be carefully controlled; microwave, infrared and optical radiation can all produce localized heating. The second problem associated with the leads into the cryostat is that they provide possible paths for significant heat leaks. Coaxial cables, waveguides, light-pipes, and direct mechanical links to rotate samples or to apply stress to them, are likewise all potential sources of undesired heat input.

Connections to experimental samples are only one aspect of the global problem of heat leaks. A basic objective of cryostat design is to ensure that the heat removed so laboriously from the sample does not quickly return again by some other path. In quantitative terms, the cooling power provided by the refrigeration system must be equal to the total heat leak into the cryostat once thermal equilibrium has been reached and the temperature is constant. Figure 7.1 illustrates this simple but fundamental principle, and in addition summarizes the most important cooling techniques together with a compendium of heat leaks. We will discuss some of the most significant of these and indicate the practical means that are most successful in minimizing them. Before beginning the detailed analysis of heat leakage for a particular design, it is important to have a firm idea of the magnitude of the available cooling power. Thus a heat input of even 1 W, a modest amount by room-temperature standards, to a ^{4}He evaporation cryostat at 4.2 K, would require a cooling power equivalent to boiling off 1.4 litres of liquid helium per hour, which is an unacceptably large amount in most experiments.

In many cryogenic systems, the lowest temperature that can be achieved is directly limited by the heat leak. For example, it is easy to show that the temperature, T, attained in a ^{4}He evaporation cryostat pumped below 4.2 K is given by

$$\dot{Q} = \frac{PL(T)}{RT} S_P \quad \text{W} \tag{7.1}$$

where $\dot{Q}$ is the heat leak, P the vapour pressure, R the gas constant, $L(T)$ the latent heat of vaporization at temperature T, and S_P the speed of the pumping system at the cryostat. For a typical system, the cooling power, and hence $\dot{Q}$, are a few mW at 1 K. Dilution refrigerators have cooling powers that are typically of the order of microwatts at millikelvin temperatures. Figure 7.2 shows the measured cooling power, as a function of temperature for a dilution refrigerator designed and built by Bradley et al. (1982). At still lower

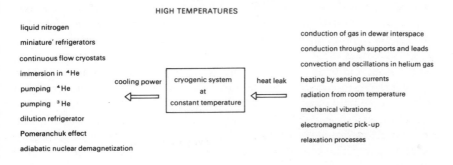

Figure 7.1   For a cryostat at constant temperature the heat leak is equal to the cooling power.

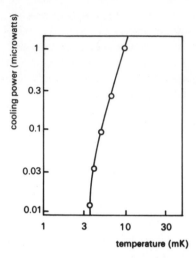

Figure 7.2 The measured cooling power of a recent ^{3}He–^{4}He dilution refrigerator. (After Bradley *et al.*, 1982.)

temperatures, the cooling power of their adiabatic nuclear demagnetization stage was estimated to be tens of pW. Simply to identify and eliminate heat leaks down to this level is a major research project in its own right.

Other points that have always to be borne in mind at the design stage are that large temperature gradients are likely to be present in the cryostat, and in addition, that many physical properties depend strongly on temperature below 300 K. The effect of thermal contraction, for instance, may bend tubes, break solder joints and wires, and seize up moving parts. Thoughtful design avoids situations in which differential contraction might occur, for example, in pairs of parallel tubes which are made of different materials, or metal tubes which pass through nylon bushes. Metals contract by between 0.1 and 0.5% between room temperature and 4.2 K, and polymers in general by about ten times that amount. It is therefore better to braze joints between dissimilar metals than to soft-solder them and no reliance can be placed on joints that are merely cemented. Fortunately, the mechanical properties of most materials—yield stress, fracture stress, fatigue properties, and creep—are all improved at low temperatures so that designs based on room-temperature values leave a safe margin for error. The noteworthy exception is that of body-centred cubic structures, such as iron and mild steels, for which the brittle strength is greatly reduced at low temperatures. An austenitic stainless steel with a relatively high chromium and nickel content should be used instead. Rosenberg (1963) gives a more detailed survey of mechanical properties at low temperatures.

As we saw in chapter 2, the heat capacity of all solids decreases with decreasing temperature and, at 4.2 K, C_V for a typical solid will be between

10^{-3} and 10^{-4} of the value at room temperature (Fig. 2.4). There are some important practical consequences that should be noted. The thermal response time of the cold end of the cryostat will be greatly reduced, so that thermal equilibrium is established very much faster than at room temperature. (This follows from the increase in thermal diffusivity $\kappa/C_V\rho$, where κ and ρ are the thermal conductivity and density respectively, both having approximately the same value at 4.2 K as at room temperature.) The very low heat capacity at low temperatures also means that only small cooling powers are needed in order to further decrease the temperature of a sample, say in a dilution refrigerator. However, at these extremely low temperatures other contributions to the heat capacity may become significant. Impurities in a crystal, for example, may give rise to a *Schottky anomaly* (see Rosenberg, 1963) and in non-crystalline materials such as nylon, epoxy, or varnish the two-level systems discussed in §2.6 will give rise to a heat capacity between 10 and 1000 times the lattice value.

Changes in electrical conductivity at low temperatures are of little significance in cryostat design. Usually leads are made from off-the-shelf, impure, polycrystalline wire whose resistivity ratio, even if the wire is of an element such as copper, is typically 10–100. In alloys such as constantan (copper 60: nickel 40), which are often used instead of copper for leads because of their much lower thermal conductivity, the resistance changes by only a few per cent. Because of the need to minimize heat leaks, the most significant physical property in cryostat design is thermal conductivity. In Figs. 2.6 and 3.9 we showed respectively the temperature variation of κ for some typical insulators and metals. The high concentration alloys such as stainless steel and cupro-nickel are most commonly used as supporting tubes or rods, because of their poor thermal conductivity. On the other hand, copper or brass specimen holders are preferable in the interests of rapid thermal equilibration and good thermal contact between different components.

Most cryostats are built on the suspension principle, as shown in Fig. 7.3, since for a given weight of cryostat the supporting tubes or rods may be made thinner if they are in tension than if they are in compression. Indeed, nylon fibres are often used. The cryostat can then be surrounded by a system of dewars containing liquid helium and liquid nitrogen in order to minimize conduction and radiation from the sides. Glass dewars are indicated in Fig. 7.3 but metal ones are equally common. White (1979) has given an expression for the heat flow, $\dot{Q}$, by gas conduction between the walls of a dewar in the limit of low pressure where the mean free path of the gas molecules is large compared with the wall separation,

$$\dot{Q} = \text{constant} \times a_0 P \Delta T \text{ W m}^{-2} \qquad (7.2)$$

where P is the pressure, ΔT the temperature difference, the constant has values of 2.1 and 1.2 for helium and air respectively, and a_0 is related to the accommodation coefficients of the surfaces, typically having a value $c.0.3$. Glass dewars are traditionally evacuated to 10^{-6} torr whilst being baked at

H

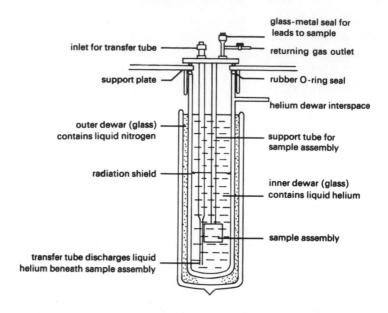

glass-metal seal for
leads to sample

inlet for transfer tube

returning gas outlet

support plate

rubber O-ring seal

helium dewar interspace

outer dewar (glass)
contains liquid nitrogen

support tube for
sample assembly

radiation shield

inner dewar (glass)
contains liquid helium

sample assembly

transfer tube discharges liquid
helium beneath sample assembly

(supporting wire cradles for the two dewars are not shown)

Figure 7.3   A typical ^{4}He immersion cryostat supported from the top.

400°C. The magnitude of the heat leak from 77 K (liquid nitrogen) to 4.2 K is then about $10\,\text{mW}\,\text{m}^{-2}$, which is low enough for most purposes. A problem with glass dewars, however, is that they are slightly porous to helium gas at room temperature, and the inner dewar of Fig. 7.3 is therefore fitted with a connection to its interspace in order to allow regular re-evacuation. Indeed, in the precooling stages of an experiment it is common practice to introduce a little air as exchange gas in order to speed the cooling process. Subsequent transfer of liquid helium into the cryostat effectively cryopumps (solidifies) the air in the interspace. Because of the porosity problem with glass, systems which require a minimal heat leak should be designed with metal dewars.

The outer dewar containing liquid nitrogen is essential in order to limit the heat leak into the cryostat by radiation. Stefan's law applied to two plane parallel surfaces of emissivity ε_1 and ε_2 for infrared radiation and at temperatures respectively of T_1 and T_2, yields an expression for the heat transfer

$$\dot{Q} = \sigma(T_1^4 - T_2^4)\frac{\varepsilon_1\varepsilon_2}{\varepsilon_1 + \varepsilon_2 - \varepsilon_1\varepsilon_2}\,\text{Wm}^{-2} \tag{7.3}$$

where σ is Stefan's constant, having the value $5.7 \times 10^{-8}\,\text{Wm}^{-2}\,\text{K}^{-4}$. The emissivity of a polished metal surface or film may be as low as 0.01 (*American Institute of Physics Handbook*, 1963) so that for $T_1 = 300\,\text{K}$ and $T_2 = 4\,\text{K}$, $\dot{Q}$

will be about $5\,Wm^{-2}$ under optimum conditions. Interposition of the dewar containing liquid nitrogen or any other surface at 77 K reduces the heat leak to $20\,mW\,m^{-2}$. Room-temperature radiation could also reach the cold end from the top of the cryostat directly down the open end of the dewars, and it is important to fit polished radiation shields, preferably anchored thermally to the dewar walls at temperatures around 50–100 K. The presence of the radiation shields also helps to inhibit convective oscillations in the helium gas above the liquid which, driven by the temperature gradient down the cryostat, can transfer large quantities of heat to the cold part of the cryostat (Keesom, 1942).

Thermal conduction down the support tubes and connecting leads, as already noted, is inevitably a major source of heat leakage. The heat flow along a metal support, of length L, assumed to be of uniform cross-section, having its ends at two different temperatures, T_1 and T_2 is

$$\dot{Q} = \frac{l}{L} \int_{T_1}^{T_2} \kappa(T)dT \tag{7.4}$$

where $\kappa(T)$ is the temperature-dependent thermal conductivity of the support. For several materials of likely cryogenic application, White (1979, page 133), has prepared a table of definite integrals with T_1 and T_2 taken as the most common pairs of temperatures encountered in low-temperature physics. For example, the heat input to a 4.2 K cryostat down stainless-steel support tubes of length $\sim 1\,m$ and total cross-sectional area $\sim 10^{-5}\,m^2$ is a few tens of milliwatts. Because of the high conductivity of copper, about 20 times that of stainless steel, sensor leads must be of as fine a gauge of wire as possible. On the other hand, if the leads are too fine, Joule heating can become troublesome. A related point is that leads from higher temperatures should always be thermally anchored before reaching the sample; otherwise local heating due to conduction may lead to unreliable measurements.

At mK temperatures when only microwatts of cooling power are available, in addition to meeting all the considerations outlined above, some other special precautions have to be taken to eliminate heat leaks. As well as having built-in infrared shields and traps, the cryostat must also be protected from longer-wavelength radiation, notably radio and television transmissions, and from emission due to other equipment in the vicinity. A common method is to enclose the dilution refrigerator assembly, together with essential support equipment, in a large Faraday cage, that is, a totally enclosed metal box known as a shielded room. Electricity supplies into the room are very carefully filtered. Heat leaks as large as microwatts can also be introduced to the cryostat through vibrations caused by rotary pumps or even by the movement of people. Millidegree practitioners therefore go to very great lengths to isolate the lowest-temperature enclosure—Bradley et al. (1982), for example, supported their cryostat on a massive concrete block floating on air springs, with foundations built directly into the clay beneath the laboratory.

For more specific details regarding all aspects of cryostat design at millidegree temperatures, the books by Lounasmaa (1974) and Betts (1976) should be consulted. The principal text on general low-temperature techniques is the book, already mentioned, by White (1979). Much useful information is contained also in the books by Hoare *et al.* (1961) and by Rose–Innes (1973).

7.2 Cooling with ^{4}He

Having reviewed the general principles of cryostat design in the previous section, we shall now describe in greater detail some specific systems for carrying out investigations at temperatures above 1 K. Cryostats for use in this temperature range are most commonly designed to obtain their cooling power from liquid ^{4}He, with pre-cooling by liquid nitrogen. It is important to bear in mind that the available cooling power of liquid helium is derived to a much larger extent from the change in enthalpy of the cold gas between 4.2 K and, say 77 K (46 J from the gas evaporated from 1 cm^3 of the liquid at 4.2 K), than from the latent heat of the liquid-gas phase transition (2.5 J cm^{-3}).

In the ideal cryostat, the temperature of the helium gas leaving the system should be as close as possible to 300 K, so that the maximum cooling power has been extracted from the liquid and cold gas. Thus, efficient heat exchange within the cryostat is very important, not only during the transfer of liquid from the storage vessel to the dewars, but also after the liquid has collected in the cold space. For example, the cold gas while leaving the cryostat must first cool radiation shields and then maintain them at a constant low temperature.

The most straightforward type of helium cryostat is that in which the sample is mounted directly in liquid helium contained in a double dewar system, as shown in Fig. 7.3. Although the heat leak into such a cryostat is relatively large because of the open top, the configuration is very convenient and easy to use. This simple design is appropriate for use in teaching laboratories and for experiments in which the low temperature is required merely to reduce thermal effects, such as noise, scattering or linewidth. The temperature of the whole ^{4}He bath is lowered by reducing the vapour pressure above the liquid and, even with a modest pump (speed $S_p \simeq 100$ litre/minute), a temperature of 1.5 K is readily obtained. Because of the direct contact with the liquid, power dissipation in components is not a significant problem and thermal equilibration is relatively fast.

When temperatures above 4.2 K are required, the sample must be (partially) thermally isolated from the helium bath by enclosing it in an evacuated can (Fig. 7.4a). The sample, which is usually mounted in a copper block with a heater and a thermometer, can then be stabilized at a temperature above that of the bath by making the thermometer control the current in the heater through a feedback loop. A simple cover, such as a layer of kitchen foil, is

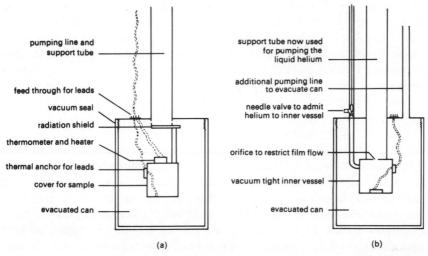

Figure 7.4 (*a*) A cryostat for measurements at stabilized temperatures above 4.2 K. (*b*) A cryostat for measurements between 4.2 K and 1 K. Liquid ^{4}He is admitted to the inner can through the needle valve, and the volume pumped through a small orifice in order to restrict superfluid film flow.

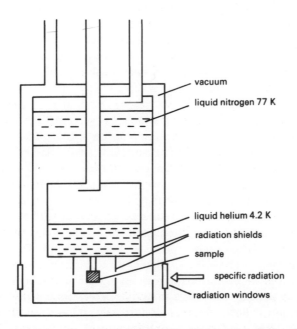

Figure 7.5 A window cryostat to allow direct access of light, with radiation shields instead of dewars.

required to prevent radiative heat transfer between the sample and its surroundings. Electrical leads are thermally anchored to the copper block before being connected to the relevant sensors.

A similar arrangement can be used to make measurements at stabilized temperatures below 4.2 K. The sample in its mounting holder is now immersed (or attached to) a separate helium container which is isolated from the main ^{4}He reservoir by a vacuum space. Liquid which is introduced into this 'can', through a needle-valve controlled from room temperature, may be cooled to 1.27 K by reducing the vapour pressure to about 1 torr. Again the temperature may be stabilized by the use of a feedback system to supply power to the heater. To obtain pressures and temperatures much below these values, careful attention must be paid to the size and shape of the orifice connecting the inner helium volume to the pumping line. As was described in §5.6, at temperatures below the λ-point of ^{4}He the inner surfaces of any containment vessel are covered with a thin film of liquid which provides an additional heat leak that becomes increasingly significant at the lowest temperatures. The film flow rate, and hence the heat leak and the rate at which helium must be removed by the pump, can be minimized by introducing a constriction in the pumping line at the entrance to the 1 K can. Temperatures down to 1 K, or even lower, can readily be achieved by use of a large rotary pump and an orifice of 0.5 mm diameter. Of course, the effect of the orifice increases considerably the time taken to pump down to the minimum temperature.

In all the systems described so far, the connections to the sample are limited to voltage and current leads or coaxial cables fed down from the top of the apparatus. There are many experiments, however, in which a more direct access for specific radiation is required in, for example, the visible, infrared, or X-ray regions of the spectrum.

In most cryostats designed to satisfy this requirement, in place of dewars, the sample assembly is surrounded by a series of cooled metal radiation shields incorporating 'windows' to allow passage of the appropriate radiation. The principle is illustrated in Fig. 7.5. The sample is suspended from the lowest-temperature liquid bath, an arrangement which lends itself to easy control of the sample temperature by the heater feedback method mentioned above. The radiation shields are usually of polished copper and do not need to be vacuum-sealed, so that the specific radiation can pass through small holes. Thus only a single pair of vacuum-tight windows needs to be provided, and these are at room temperature. Allowance must be made for the fact that room-temperature thermal radiation will also be reaching the sample through the windows, and it is important to provide very good thermal contact with the helium bath. A variety of window materials are commercially available for different radiations. Some common ones are sapphire for ultraviolet and visible radiation, Mylar or rocksalt for infrared, and beryllium or Mylar for X-rays and neutrons. Another method of bringing radiation into a cryostat is by means of a light guide or optical fibre, but limitations on wavelength are imposed by the reflectivity of the guide or the transmission characteristics of the fibre. The heat leak, however, is not excessive.

Less common experimental studies may require additional ingenuity in cryostat design. Magnetic susceptibility measurements by a static technique, for example, pose particular problems. In the Faraday method the force on the sample in an inhomogeneous magnetic field is measured by a balance at room temperature connected to the sample at low temperature by a glass fibre, as described by Zilstra (1969). Cryogenic problems arise because any direct contact with the sample will affect the measurement, whilst on the other hand room-temperature radiation passing down the tube surrounding the fibre will be directly incident on the sample. Hedgecock and Muir (1960) describe a cryostat which overcomes some of the difficulties.

A fairly common requirement is for the application of uniaxial stress or hydrostatic pressure to a sample at low temperatures, calling for a direct mechanical link between top and bottom of a cryostat. As we have noted, solids have greater strength in tension than in compression; and tubes, or even wires, may be used to transmit the force to the 'anvil' holding the sample. It is crucial to ensure that the system of forces does not exert a torque on the cryostat. There are of course no liquids available, other than helium, for hydrostatic experiments below about 20 K, and ^{4}He itself solidifies at 25 atmospheres below 1 K. Solid helium, however, has very low shear strength and can be used for approximately hydrostatic transmission of pressure up to around 10^4 atmospheres. For further details, see Swenson (1960), and Webb et al. (1976).

Many cryogenic devices, although requiring low temperatures in order to function, are small in size and dissipate very little power in operation. Examples are low-noise amplifiers, SQUIDS and other Josephson devices (described in §8.3), and radiation detectors and bolometers. The localized cooling power required may be less than a milliwatt whereas the liquid ^{4}He cryostats in which they are commonly immersed can supply cooling powers 1000 times larger. The trend in recent years has been towards closed-cycle cooling systems (refrigerators) which provide refrigeration without direct immersion in liquid helium. Continuous-flow cryostats usually derive their cooling power from liquid helium piped from a transfer dewar directly on to the sample chamber. The reverse flow of vaporized ^{4}He keeps the insulated inlet pipe cold through a carefully-designed heat exchanger, and the flow rate determines the lowest temperature reached by the sample. Details of a typical continuous-flow design have been given by Campbell et al. (1976). An alternative approach to the problem is through 'microminiature' refrigerators, reviewed recently by Little (1982). Based on the Joule–Kelvin effect, these devices have no moving parts except for the remote compressors. Some idea of their size can be obtained from Fig. 7.6 (p. 202) which shows the planar, glass Joule–Kelvin element, and the complete refrigerator. The very fine capillaries are about 63 μm wide and are produced by photo-lithography. So

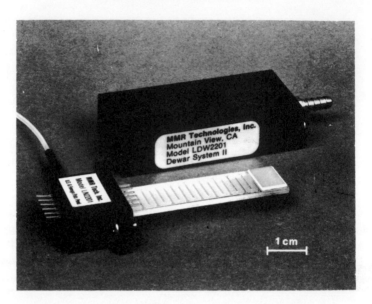

Figure 7.6   A 'microminiature' Joule–Kelvin refrigerator capable of reaching 83 K. (After Little, 1982). Reproduced by courtesy of the author.

far only relatively high temperatures have been achieved, using nitrogen gas as the working fluid, but work is in progress to reach helium temperatures using a three-stage process involving nitrogen, hydrogen and helium.

7.3 Cooling with ^{3}He

In this section we will look at three different ways in which the unique properties of ^{3}He are used for obtaining temperatures below 1 K; evaporation cooling of the liquid, Pomeranchuk cooling by adiabatic solidification of the liquid, and dilution cooling by mixing ^{3}He with ^{4}He.

^{3}He evaporation cryostats

Helium-3 is not liquefied in bulk by the methods outlined in chapter 1; instead, a small amount of liquid is condensed in the cryostat itself by bringing gas into thermal contact with a ^{4}He bath which has been cooled well below the critical temperature (3.35 K) of ^{3}He. The usual condensation temperature will be about 1.2 K and the molar enthalpy change, as the liquid is cooled to a base temperature of 0.3 K at a vapour pressure of 1.9 torr, is 3.7 J. As the average latent heat over this temperature range is about 34 J mole^{-1}, there is only a 10% boil-off as the liquid is cooled, if other losses are ignored. A schematic

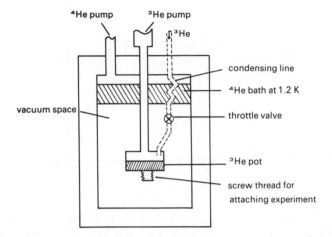

Figure 7.7 Schematic diagram of a ^{3}He evaporation cryostat. The dashed line is included to make a continuous cycle refrigerator..

diagram of the ^{3}He stage of a cryostat is shown in Fig. 7.7. A few ml of liquid are condensed into a small container (pot), usually made of copper, which is mounted in a vacuum can. Rather than being directly immersed in the liquid, the sample which is to be cooled is usually attached to the outside of the pot using a low-melting-point solder (such as Wood's metal), or by a screw. In a single-cycle cryostat, the ^{3}He which is exhausted from the pumping system is returned to the store and refrigeration finally ceases when no liquid remains in the pot. However, provided that the heat leak from all sources is 50 μW or less, a base temperature of below 0.4 K can be maintained for about 3 hours starting with 1 ml of liquid. If greater cooling power is required and the base temperature is less important, a ^{3}He cryostat can be operated as a cont-inuously working refrigerator by feeding back the exhaust gas into the ^{3}He pot via a suitable condensing line (dashed line in diagram). The rate of feed-back is controlled by a throttle which may be either an adjustable valve or a fine capillary tube.

Since the lowest temperature that can be obtained with an evaporation cryostat is limited by the rate at which vapour can be pumped off the liquid, some care is needed in the choice of dimensions for the pumping lines between the ^{3}He pot and the diffusion pump. Very high pumping speeds, and hence low temperatures, may be conveniently achieved by use of an activated charcoal sorption pump: this is connected to the ^{3}He pot via a short length of line and is cooled by the ^{4}He bath. Although a base temperature of 0.28 K with a heat load of 50 μW was reached by Mate et al. (1965), the lowest temperature reported ($T \sim 0.21$ K) was achieved by use of a carefully-designed pumping system coupled to a large diffusion pump (Walton, 1966).

Pomeranchuk cooling

In chapter 6, the properties of ^{3}He in the vicinity of the melting curve were discussed. Reference to Fig. 6.6 shows that the deep minimum in the melting curve at $T = 0.32$ K and $P = 29$ atm is associated with the cross-over in the temperature-dependent entropies of the liquid and solid phases. Pomeranchuk (1950) proposed that cooling of ^{3}He would result from the adiabatic compression of a liquid-solid mixture along the melting curve (A'B' in Fig. 6.6), with the lower temperature limit of about 1 mK being determined by the onset of anti-ferromagnetic order in the solid. Anufriyev (1965) was the first to demonstrate the method and, since then, a number of low-temperature laboratories around the world have developed the technique to achieve temperatures close to 2 mK. The basic method can be stated very simply. By suitable precooling (by means of a dilution refrigerator or by an adiabatic demagnetization stage) and application of pressure, liquid ^{3}He in a cell is brought to the point A' on the melting curve in Fig. 6.6. Increasing the pressure in the cell produces solidification and thence Pomeranchuk cooling, represented by the horizontal line AB on the entropy temperature curve, where the point B corresponds to the cell containing just solid. However, if the starting temperature is sufficiently low, the fraction of the solid in the cell near 3 mK may be quite small. For example, if $T_i = 25$ mK and $T_f = 3$ mK, only 20% of the original ^{3}He will solidify during the compression, and the remainder is available for cooling a sample. The cooling power of the method is simply the amount of heat that can be absorbed when n moles/second of liquid are converted into solid at constant temperature T

i.e. $$\dot{Q} = T(S_s - S_l)\dot{n}.$$ (7.5)

As the entropy difference is approximately constant over most of the usual working range of refrigeration, $\dot{Q}$ is proportional to absolute temperature. A quantitative comparison with a dilution refrigerator is given below.

The main experimental problem with Pomeranchuk cooling is to devise a method for compressing the helium which avoids frictional heating. This restriction usually rules out mechanical means which would anyway be cumbersome to operate at very low temperatures. Any attempt to force ^{3}He into the cell by compression from room temperature would be totally ineffective because just below T_{min} a plug of solid ^{3}He forms in the filling tube and prevents further compression. External hydraulic compression of the cell with liquid ^{4}He resolves this problem but, since ^{4}He itself solidifies at this temperature at a pressure P_s of less than P_{min}, it is essential to use some method of amplifying its pressure without exceeding P_s (Lounasmaa, 1974; Betts, 1976).

Pomeranchuk cooling provides an ideal method for studying the properties of ^{3}He, albeit only along the melting curve, and it was the technique being used in the experiments in 1972 (see §6.4) that led to the discovery of new phases of

^{3}He. The high cooling power is an attractive feature but, as interest in the properties of matter below 1 mK has increased, the more versatile alternative of a two-stage refrigeration system consisting of a dilution refrigerator coupled to a nuclear cooling stage has become the standard method.

$^3He-^4He$ dilution refrigerators

Although single refrigerant cryostats using liquid ^{4}He and ^{3}He are limited to base temperatures of about 1 K and 0.3 K respectively, a technique which employs solutions of ^{3}He in ^{4}He is capable of maintaining continuously temperatures as low as 2 mK. The method depends on another property discussed in chapter 6, namely that of phase separation of ^{3}He–^{4}He mixtures below 0.9 K. If a mixture with a concentration of ^{3}He greater than 6% is cooled from a temperature of about 1 K (represented by the point X in Fig. 6.14), then phase separation occurs at the point A into a ^{3}He-rich (concentrated) phase which floats on top of the more dense ^{4}He-rich (dilute) phase corresponding to A′. Further cooling along the co-existence curve causes the concentrated phase to become almost entirely 100% pure ^{3}He, while the dilute phase tends to a limit of 6% ^{3}He in superfluid ^{4}He. It is this finite zero-temperature ^{3}He concentration in the dilute phase that makes possible the process of dilution cooling. When ^{3}He atoms move across the boundary from the concentrated to the dilute phase, a heat of solution is taken from the liquids. This process is analogous to the evaporation of a liquid in a conventional cryostat (§7.2) in which the extraction of latent heat produces cooling. As already discussed in §6.5, ^{3}He atoms in the dilute superfluid phase behave like the particles of a gas, moving through the inert ^{4}He background as if it were a vacuum, and at a 'vapour' pressure which is the osmotic pressure for a solution of ^{3}He in ^{4}He. In order to maintain a continuous flow of ^{3}He atoms across the phase boundary, an osmotic-pressure gradient must be established in the dilute solution. This is achieved if the mixing chamber, wherein the phase separation occurs, is connected through a tube to a second chamber, called the still, which is maintained at a temperature of about 0.7 K (see Fig. 7.8). When the liquid in the still is pumped, the vapour which is removed is almost entirely ^{3}He because its vapour pressure is so much higher than that of ^{4}He (the ^{4}He film flow being restricted by a suitable orifice in the pumping line exit), and the required flow of ^{3}He for cooling the mixing chamber is thus established.

As with a ^{3}He cryostat, the dilution process can be operated as a single cycle. Once the phase separation has been established, the concentrated phase in the mixing chamber is continuously depleted by removal of ^{3}He and the cycle is completed when no more ^{3}He remains. Most systems, however, operate as continuous dilution refrigerators with the evaporated ^{3}He being recirculated back to the mixing chamber. The principal components of the 'conventional' continuous cycle refrigerator are shown in Fig. 7.8. As with a ^{3}He refrigerator, the returning gas is condensed by thermal contact with a ^{4}He pot at about

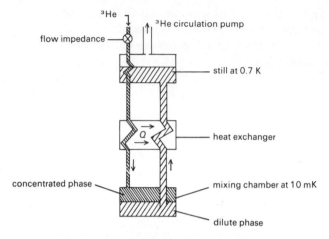

³He

flow impedance

³He circulation pump

still at 0.7 K

Q

heat exchanger

concentrated phase

mixing chamber at 10 mK

dilute phase

Figure 7.8   Schematic diagram of a ³He/⁴He dilution refrigerator.

1.2 K and the liquid then passes through a sequence of heat exchangers in which cooling by contact with the counterflowing dilute phase occurs. Design of these heat exchangers is then the key to the successful operation of the dilution refrigerator and is a crucial factor in minimizing the base temperature and in maximizing the cooling power. However for a simple system a single concentric-tube heat exchanger is sufficient to provide a base temperature of 50 mK (Anderson, 1970).

In chapter 1 we noted that enthalpy was the appropriate thermodynamic potential for an analysis of the Joule–Kelvin refrigerator which incorporated a number of heat exchangers. The same quantity is used in the analysis of a dilution refrigerator and its cooling power is determined by the difference in the partial enthalpy of ³He between the dilute and concentrated phases. For a mixing chamber at a constant temperature T_m, it can be shown (Lounasmaa, 1974) that for a dissolution rate of $\dot{n}$ moles/second, the cooling power $\dot{Q}$ is given by

$$\dot{Q} = 84\dot{n}T_m^2\,\mathrm{W} \qquad (7.6)$$

for a single-cycle refrigerator.

The T^2 term arises through the integration of the linearly temperature-dependent specific heat of ³He (§6.2) in both phases. In the continuously operated refrigerator $\dot{Q}$, of course, is reduced by the enthalpy of the returning ³He liquid which leaves the final heat exchanger at $T > T_m$.

Some comparisons between the ³He cooling methods can now be given. The advantage of the more complex dilution refrigerator over the simple ³He evaporation cryostat becomes increasing apparent below 0.5 K. For a room-temperature pump which has a constant speed, that is, volume of gas handled/second, the rate of removal of ³He atoms by evaporation $\dot{n}$ is proportional to the vapour pressure which decreases exponentially with temperature (see §7.5) while the heat removed per mole remains approximately constant at L, the latent heat. In the dilution refrigerator, $\dot{n}$ is approximately constant below 0.2 K, but the cooling capacity per mole falls as T_m^2. The exact temperature at which the exponentially decreasing cooling power of the ^{3}He cryostat falls below that of the dilution refrigerator is dependent on the pumping speed but for a typical system, the cross-over point occurs at about 0.35 K. The cooling power of the Pomeranchuk refrigerator, which is given by (7.5), is proportional to the absolute temperature over the temperature range in which $\Delta S = S_s - S_l$ is approximately constant. For a circulation rate $\dot{n}$ equal to the molar rate of solidification, the cooling power of the dilution refrigerator at 10 mK is a factor 10 smaller than that of a Pomeranchuk cryostat.

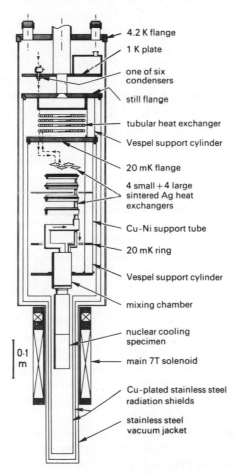

Figure 7.9 Schematic view of a Lancaster dilution refrigerator used as the precooling stage in a nuclear cooling cryostat. (After Bradley *et al.*, 1982).

Many successful designs of dilution refrigerators have been described in the literature, and commercial systems are available in both the U.K. and the U.S.A. Lounasmaa (1974) provides detailed information about some complete systems and their component parts. Here, we shall give as an example a brief account of the dilution refrigerator which has been built at Lancaster University for use with a nuclear cooling stage (Bradley et al., 1982). Its important feature is high cooling power at a low base temperature; two conditions which must both be satisfied for successful nuclear cooling. An outline diagram of the main components is shown in Fig. 7.9. The base temperature depends on the heat carried by the returning ^{3}He to the mixing chamber as well as on the external heat leak. In order to minimize the former heat flow, the design of the bottom heat exchangers is crucial. Contact between the cold dilute stream and the warmer incoming liquid is limited by the Kapitza boundary resistance (§2.7) and, in the Lancaster refrigerator, the 8 discrete exchangers below the 20 mK plate use sintered silver powder to provide the largest possible area of contact between helium and the body of the exchanger. Flow channel dimensions are graded in size to reduce the effects of viscous heating which become more serious as the temperature is reduced, while the lengths of the channels are calculated so as to minimize longitudinal conduction. The minimization of external heat leaks into the mixing chamber requires as much attention to detail as does the design of cryogenic components. All of the factors which are listed in §7.1 contribute to the heat leak but it is important to pay particular attention to those which can usually be neglected in higher-temperature cryostats; for example, vibrational heating from sources within and without the cryostat, and induced electrical heating in metal parts which originates from unscreened external r.f. sources. A base temperature of less than 3 mK is achieved by the Lancaster refrigerator and with a circulation rate for ^{3}He of 150 μ mole sec^{-1}, the cooling power at 10 mK is 1 μW, although it falls to about 0.01 μW at 3 mK, as shown in Fig. 7.2.

7.4 Magnetic cooling

The principles of magnetic cooling, or adiabatic demagnetization, apply both to paramagnetic spins in salts and to the nuclear spins in metals although the experimental techniques used to achieve the low temperatures are somewhat different. Even though a complete thermal cycle may last hours or days, magnetic cooling is a single-cycle refrigeration process and, once the adiabatic stage has started, the temperature of the magnetic system and any thermal load attached to it are likely to be changing continuously. This is one of the reasons why cooling with a paramagnetic salt has largely been replaced by continuous-cycle dilution refrigeration when temperatures in the range from 0.3 K down to a few mK are required. For sub-mK temperatures, however, nuclear cooling is the only available option.

As mentioned above in §1.2, the application of a magnetic field to an assembly of N weakly interacting magnetic dipoles, either electronic or nuclear, results in a reduction in entropy since the magnetized state has greater order. If we have a single ion with total ground-state angular momentum J, there is an equal probability to zero field of it occupying any one of the $(2J + 1)$ degenerate eigenstates and the entropy of the whole assembly of ions is $Nk_B \ln(2J + 1)$. In a real crystal the ground state cannot be truly degenerate, since interactions between individual dipoles and the lattice and between each other will cause internal energy splittings whose average value we will call ε_i. Thus the expression for S is true only for $k_B T \gg \varepsilon_i$. Since the entropy of the magnetic system must tend to zero at $T = 0$ in accordance with the Third Law, an ordering process will set in as the temperature is reduced and all the ions will eventually occupy a singlet ground state. In a finite magnetic field at $T > 0\,$K, the occupation of the levels, which are separated by the Zeeman energy, will follow a Boltzmann distribution. For a simple model of non-interacting spins, both the entropy and the magnetization of the dipole system are functions of (B/T) where, to a good approximation, B is the magnitude of the externally applied field.* Figure 7.10a is a representation of the entropy versus temperature variation in zero and finite magnetic field for the assembly of either atomic or nuclear spins. This discussion has considered only the entropy of the spin system and the total entropy of the solid will include, of course, lattice and (possibly) conduction electron components. Although we may assume that, for the given starting conditions ($T \simeq 1\,$K for paramagnetic salt cooling or $T \simeq 20\,$mK for nuclear cooling), these additional contributions are small, Fig. 7.10a strictly represents the magnetic entropy.

The magnetic field is applied under isothermal conditions at a starting temperature T_i. During this stage of the cycle (X–Y) the heat of magnetization must be removed by maintaining the spin system in good thermal contact with a constant-temperature bath; for example, with a 1 K ^{4}He bath or with the mixing chamber of a dilution refrigerator. When steady-state conditions prevail, the thermal switch is opened to isolate the magnetic stage (refrigerant plus sample) and the magnetic field is reduced from its initial value B_i under adiabatic conditions. Isentropic cooling of the spin system is achieved in the stage Y–Z. If we assume that heat leaks into the system commence only when the final temperature has been reached in zero or small magnetic field (B_f), the warming stage follows the path Z–X or Z'–X' to complete the cycle. The temperature T_f is strictly the spin temperature of the dipole system, that is, it is the temperature in the Boltzmann factor $k_B T$ which would describe the population of the $(2J + 1)$ levels which was established in the magnetizing field B_i and which is 'frozen' during the subsequent adiabatic field reduction to B_f. In Fig. 7.10b the relative populations of the energy levels are shown for these points in the cycle. Since the magnetization of the assembly does not change

---

* We will see in the thermometry section that the difference between the local field at the site of the dipoles and the applied field depends on the sample magnetization and shape.

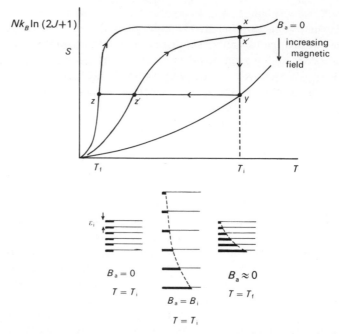

Figure 7.10 (a) Entropy-temperature diagram for a system of magnetic dipoles in three different values of the applied magnetic field. (b) Representation of the occupation of energy states during the magnetic cooling cycle.

between Y and Z—magnetic cooling involves an adiabatic reduction of magnetic field—the final temperature must be related to the starting temperature by

$$\frac{B_i}{T_i} = \frac{B_f}{T_f}. \tag{7.7}$$

We cannot, of course, achieve $T_f = 0$ by making $B_f = 0$. When $k_B T \sim \varepsilon_i$, the magnetization and entropy can no longer be expressed as a function of B/T; equation 7.7 is not valid and the final temperature is determined by the ordering process which causes the zero applied field entropy to fall to zero. If the splitting ε_i is related to an equivalent internal magnetic field b, (7.7) should be rewritten as

$$\frac{(B_i^2 + b^2)^{1/2}}{T_i} = \frac{(B_f^2 + b^2)^{1/2}}{T_f}. \tag{7.8}$$

Since $B_i \gg b$, the condition for a minimum value of T_f is that the ratio (B_i/T_i) should be a maximum. If the final field $B_f = 0$, equation 7.8 reduces to

$$T_f = T_i \frac{b}{B_i}.$$

For magnetic cooling with paramagnetic salts, the refrigerant is either a single crystal or is a powder of small crystals which is compressed into an ellipsoidal-shaped pill. During the isothermal magnetization, the salt plus sample and thermometer are maintained in thermal contact with a pumped helium bath by allowing ^{4}He, as the thermal 'exchange gas', to fill the vacuum space of Fig. 7.11a. When the heat of magnetization has been removed and the temperature stabilized, the exchange gas is pumped out so that the adiabatic reduction in magnetic field occurs with the salt isolated from its surroundings. Double sulphates of iron group elements (alums) and cerium magnesium nitrate (CMN) are the most commonly used materials but, if the lowest possible temperature is required, CMN has an internal b field which is a factor 10 smaller than the alternatives. For a starting temperature of 0.5 K and $B_i = 1$ T, a temperature of 2 mK can be reached. If a large cooling power, rather than a minimum base temperature, is the principal requirement, however, the field is reduced, not to zero, but to a final value which is greater than zero, so that the heat capacity throughout the warm-up stage remains relatively high. This result can be seen by reference to Fig. 7.10a: the amount of heat that can be absorbed during the warm-up at constant or zero magnetic field is just

$$\Delta Q = \int T dS$$

or the area under the portion of the $S - T$ curve XZ or X′Z′ and the S axis. The same effect is obtained by cooling a paramagnetic salt with a larger internal field b to zero applied magnetic field.

The nuclear moment in a metal is some 2000 times smaller than the electron moment of a paramagnetic salt, but the internal field b, if it is determined solely

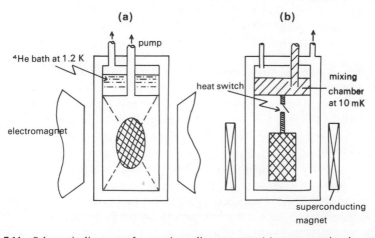

Figure 7.11   Schematic diagrams of magnetic cooling cryostats: (a) paramagnetic salt suspended in a vacuum can: thermal contact to the 1 K bath is made via exchange gas; (b) copper nuclear refrigerant: thermal contact to the mixing chamber of a dilution refrigerator is via a superconducting heat switch.

by dipole–dipole interactions, is correspondingly reduced and, in principle, spin temperatures of less than $1\,\mu$K can be reached by nuclear cooling. However the initial conditions are rather more demanding because the ratio (B_i/T_i) must be large if there is to be a significant reduction in entropy when the magnetic field is first switched on.

Several metallic elements possess the necessary thermal and magnetic properties to make them possible candidates for nuclear refrigeration and their relative merits are discussed by Lounasmaa (1974). Although copper is the most commonly used material, the choice owes as much to its ready availability in the form of wire or powder as to its other properties. Its high electronic thermal conductivity is an additional consideration. Using the values given by Lounasmaa, we shall look at the final temperature and available cooling power when a copper sample is demagnetized from an external field of 6T at a starting temperature of 10 mK. The usual practice is not to attempt an isothermal magnetization at T_i but, instead, to apply the field when the precooling stage, usually a dilution refrigerator, is above its base temperature and greater cooling power is available. At the point corresponding to Y of Fig. 7.10a in this cooling cycle, the degree of magnetic polarization is 27% but the entropy has only been reduced by 5% from its limiting value of $Nk_B \ln(2I + 1)$, where $I = 3/2$ is the nuclear spin for copper. A very low final spin temperature is still achieved if the external field is reduced adiabatically to zero; for $b = 0.3\,$mT, $T_f = 0.5\,\mu$K. If the field is reduced to 60 mT, the final spin temperature is 10^{-4} K but the heat load that can be absorbed as the nuclei return to the original temperature T_i is about 0.12 mJ/mole of copper.

Throughout the discussion so far, we have assumed that the spin temperature T_s is the same as the temperature of everything else, including the conduction electrons (in the case of a metal), lattice and any material that is to be cooled. This assumption is only valid at the points X and Y and at some late part of the warming stage Z'X' or ZX. In the intermediate region, the complete system is characterized by a number of different temperatures which are appropriate to the individual component parts. A full analysis of the internal energy exchange between these components and the effects of external heat leaks must be made in order to determine the optimum materials and conditions of operation for a magnetic cooling refrigerator. This is particularly so for nuclear refrigeration. We will illustrate the problem and its solution by referring again to the same large cryostat at the University of Lancaster which is used to cool ^{3}He to temperatures below 0.2 mK. A schematic diagram of the nuclear stage is shown in Fig. 7.12. The precooling stage is the dilution refrigerator which was described in §7.3 and thermal contact between the mixing chamber and the nuclear refrigerant, copper powder, is controlled by a superconducting heat switch which is opened or closed (§4.2) by the fringing field of the main solenoid (see also Fig. 7.12).

The thermal path between the nuclear spins and the ^{3}He liquid is illustrated by Fig. 7.13. The spin–lattice coupling is via the conduction electrons and is

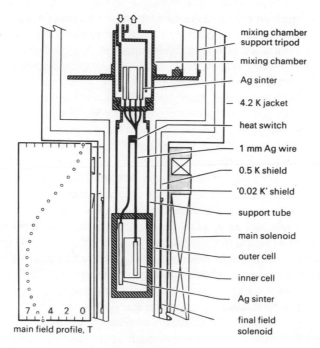

Figure 7.12   Nuclear cooling stage of a Lancaster cryostat. The inset shows the fringing field from the solenoid which is used to control the superconducting aluminium heat switch. (After Bradley *et al.*).

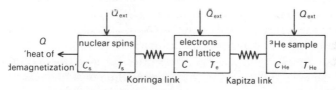

Figure 7.13   Energy flow between different thermal components in a nuclear cooling experiment. It is assumed that thermodynamic equilibrium is maintained within individual components so that each one can be described by a single temperature.

characterized by a relaxation time τ_1 given by the Korringa relation (1950)

$$\tau_1 T_e = K \tag{7.9}$$

where the quantity K is independent of temperature but is slightly field-dependent. At 10 mK, τ_1 is of the order of seconds in metals, whereas in insulators it can be as long as days or weeks; this is one of the reasons why metals are used in nuclear cooling. It is usually assumed that the copper lattice, which has a negligible heat capacity at these temperatures, is at the same temperature as the electrons, so that the combined electron-lattice system is represented by a single thermal capacity C at a temperature T_e. Cooling of the ^{3}He is via the Kapitza boundary resistance (§2.7). Unavoidable external heat leaks, which result from mechanical vibration, r.f. heating, heating produced by thermometers and thermal conduction through supports and so on, may cause energy dissipation in all three heat capacities. Coupled equations can be set up for the rates of change of T_s, T_e and T_{He} with time following the start of the field reduction, and these can then be solved with appropriate values of heat leaks, either estimated or measured. In the Lancaster cryostat, the initial field is 7T and the initial temperature is typically less than 10 mK. The field is reduced to its final value of 14 mT (a field which may be used for NMR thermometry, as described in §7.5) over a period of 3 hours. Measurements of the ^{3}He viscosity indicate that the liquid has been cooled to a base temperature of 125 μK within one hour of the end of the field reduction. Most of the nuclear cooling systems that are to be found in the literature have used copper; either in powder form to give large surface area for making thermal contact with a liquid He sample, or as a bundle of copper wires to improve thermal conduction throughout the sample while minimizing the effects of eddy current heating. The required polarization is produced by precooling to as low a temperature as possible while applying a large external field.

An alternative approach, which requires more modest temperatures and fields, is to use an enhanced hyperfine interaction as a means of generating a very large magnetic field at the site of the nucleus (Al'tshuler, 1966). Ions which have an (electronic) singlet ground state and are therefore non-magnetic in the absence of an applied field, may become magnetic by coupling to an excited state when a field is applied. This induced magnetic moment in turn produces the enhanced hyperfine field at the nucleus. A limited number of intermetallic compounds show this behaviour at low temperatures. Of these PrNi$_5$ has attracted the greatest interest because of its (relatively) low ordering temperature, $T \simeq 0.4$ mK. Starting with an initial temperature of 30 mK, where the high cooling power of a dilution refrigerator can be exploited to the full, an external field of 6T produces an entropy reduction of 50%. Thus, although the final temperature following the adiabatic field reduction is greater than that which can be reached with copper, the larger cooling power can be used to good effect in a three-stage refrigerator, in which the final temperature of the PrNi$_5$ stage is the initial temperature for a third stage with a copper refrigerant (Mueller et al., 1980).

7.5 Thermometry and instrumentation

The principles of thermometry at low temperature are essentially the same as those in any other range of temperature but there are, of course, problems associated with measurement that are peculiar to cryogenic systems and which become progressively more serious as the temperature is reduced below 1K. Any physical property which changes with temperature can provide the basis for a thermometer and it will be clear from the earlier chapters that there is no shortage of candidates for low-temperature work, even if their individual useful ranges are limited. In practice the choice of thermometric parameter is restricted by a number of prerequisites: the parameter (X) must be capable of convenient and precise measurement, it must be simply related to the absolute temperature T, and its temperature sensitivity (dX/dT) should be high: the thermometer should provide reproducible readings and have a short response time for change in temperature. Maintaining or establishing thermal contact between thermometers and test samples and minimizing the effects of the heat generated by the measurement process are just two of the potential difficulties associated with very low temperatures.

In chapter 1, we noted that the absolute or Kelvin scale of temperature is based on the quantities of heat which are transferred to and from the working substance in an ideal heat engine (Carnot cycle). The ratio of the temperatures of the two reservoirs of infinite heat capacity between which the cycle operates is equal to the ratio of the heat exchanged during the two isothermal stages, i.e.

$$\frac{T_1}{T_2} = \frac{Q_1}{Q_2}$$

and is independent of the working substance. Two fixed points define the Kelvin scale: $T_2 - 0$ (corresponding to $Q_2 = 0$) and the triple point of water $T = 273.16$ K. In principle, any unknown temperature can be determined by measuring Q_1 and Q_2 for a Carnot cycle operating between two reservoirs at the triple point and the unknown temperature respectively, so that $T = 273.16 \times Q/Q_{tp}$. In practice the idealized Carnot cycle cannot be realized, but we can use the laws of thermodynamics and the properties of ideal or nearly ideal substances to determine the absolute temperature scale. If the working substance of our Carnot cycle were an ideal gas, for which the product PV is a function of temperature only, we can show (Zemansky and Dittman, 1981, ch. 7) that the gas temperature (θ) which is defined by $PV = a\theta$, where a is a constant, is the same as the absolute temperature T. A fundamental property of real gases is that in the limit of zero pressure, the product PV, is independent of the nature of the gas and depends only on T and thus we can redefine the absolute temperature scale by

$$T = 273.16 \frac{\text{limit}(PV)_T}{\text{limit}(PV)_{tp}}$$

Gas thermometers have therefore played an important part in defining the internationally accepted temperature scales over a very wide range of temperatures to a lower limit of 2.6 K. Magnetic thermometry, based on the relationship between magnetic susceptibility and temperature for a paramagnetic salt, is used to extend the scale to lower temperatures, since for an ideal paramagnet (one that obeys Curie's law exactly), the magnetic temperature, defined by

$$T_m = \frac{\text{Curie constant}}{\text{susceptibility}},$$

is equal to the absolute temperature. A discussion of the measurement of standards and the establishment of temperature scales is inappropriate here and the reader is referred to a review by Rubin *et al.* (1982) as a starting point for detailed discussion of what is actually a very complex subject. It is worth noting, however, that the 1976 provisional scale, which was recommended by an international committee for use between 0.5 K and 30 K, is defined in terms of a number of fixed points, which include the superconducting transition temperatures of a number of pure metals, the normal boiling points of ^{4}He and H_2, and the triple point of H_2.

The thermometers that the experimental low-temperature physicist incorporates in a cryostat will fall into one of two categories, namely primary and secondary. The former is used without calibration and is often there to provide a means for calibrating a secondary thermometer at one or more known temperatures. The secondary thermometer is generally the more convenient to use, has greater sensitivity and speed of response, and may function outside the temperature range of the primary thermometer. Although the constant-volume gas thermometer in principle provides results directly in terms of the absolute temperature, it is a cumbersome instrument and a number of corrections must be made to achieve a precision of better than 10 mK: consequently it is not regularly employed as a primary thermometer in the research laboratory.

Vapour-pressure thermometry is one of the standard techniques for determining temperatures throughout the ranges that can be obtained by pumping on cryogenic liquids. The Clausius–Clapeyron relation

$$\frac{dP}{dT} = \frac{\Delta S}{\Delta V} = \frac{L}{T\Delta V}$$

may in principle be integrated to find a relationship between the equilibrium vapour pressure and temperature. If the change in volume ΔV at vaporization is taken to be the volume of an ideal gas at the same pressure and temperature, we then need to know only how the latent heat L varies with temperature. If L is a linear function of temperature

$$L = L_0 + aT$$

then

$$\ln P = -A/T + B \ln T + C$$

Experimental results fit quite well to equations of this type but the tables of P and T for ^{3}He and ^{4}He are in fact secondary scales: the 1958 ^{4}He scale (T_{58}, covering the range 1.0–5.2 K) is based on a series of gas-thermometer measurements (described by Van Dijk, 1960), and the 1962 ^{3}He scale (T_{62}, covering the range 0.3 to 3.3 K) is derived from a combination of thermodynamic calculation and an isothermal comparison of ^{3}He and ^{4}He vapour pressures (Sydoriak et al., 1964). For research purposes, the He vapour pressure tabulation provides a continuum of fixed points against which other thermometers, e.g. carbon resistors, may be conveniently calibrated.

Experimental details of both gas- and vapour-pressure thermometry are to be found in the standard texts (White, 1981; Rose-Innes, 1973). In its simplest form, the vapour-pressure thermometer consists of a pressure-measuring device (a manometer or Macleod gauge) connected to a tube inserted into the cryostat so that its lower open end is immediately above the liquid surface. In a more sophisticated version, the tube terminates in a bulb or hole within a copper block. The block is immersed in the pumped liquid bath and pure gas is condensed into the vapour pressure bulb. In addition the tube may be given a vacuum jacket to reduce effects of temperature inhomogeneities within the liquid. Three particular problems associated with low-temperature vapour-pressure thermometry should be noted, two of which result from the film flow of superfluid ^{4}He up the walls of the sensing tube. Reflux of the evaporated film carries heat back to the liquid, causing its temperature to rise above that of the bulb and block, and hence above that of the liquid bath. In addition there is a pressure drop along the sensing tube because of the viscous return flow of the evaporated film. The third problem, which also causes the measured pressure to differ from the true vapour pressure at the liquid surface, is common to any thermometer in which the pressure gauge and the volume of gas or vapour whose pressure is to be measured are not at the same temperature and are connected by a tube which has a diameter less than the mean free path of the gas molecules. However this thermomolecular pressure effect becomes important only at pressures less than 1 torr if the tube diameter is greater than a few millimetres.

Electrical thermometers have many of the required features mentioned at the beginning of this section. Their main drawback is that the thermometric parameter, usually the resistivity (thermocouple thermometry is discussed below) seldom follows a simple temperature dependence in many materials at low temperatures and may also be a function of magnetic field. Pure metals have a linear resistivity–temperature relationship at room temperature, but their cryogenic behaviour is much more complicated, since the resistivity is a function of sample purity and varies considerably between different materials. According to Matthiesen's rule (§3.3), the low-temperature resistance may be written as

$$\rho(T) = \rho_{imp} + \rho_1(T)$$

where ρ_{imp} is the temperature-independent, residual resistivity arising from impurity scattering of conduction electrons, and $\rho_1(T)$ is the temperature-dependent resistivity which arises from phonon scattering ($\rho \propto T^n$ where

$3 < n < 5$) and possibly from electron–electron scattering ($\rho \propto T^2$). Although platinum resistance thermometers play an essential role in the realization of international temperature scales above the boiling point of O_2 and have good reproducibility at all temperatures, the usefulness of metallic resistance thermometers below 10 K is limited because both ρ and $d\rho/dT$ become small.

It might be expected that metallic alloys would make very poor thermo-metric materials. In §3.4, however, we saw that scattering of conduction electrons by magnetic impurities (the Kondo effect) leads to a minimum in the low-temperature electrical resistance of some alloys. The increasing resistance of both **CuFe** and **AuFe** alloys below the minimum at about 20 K provides in principle a useful extension to the range of resistance thermometers, but as temperature indicators their value is limited by a fairly low sensitivity and by uncertainty in their degree of reproducibility.

Whereas the resistivity of a metal becomes very small at low temperatures, that of a semiconductor increases as the temperature is reduced. The resistivity, however, is not a simple function of temperature unless the impurity level is so low that the material behaves like an intrinsic semiconductor having an energy gap E_g, i.e.

$$\rho \propto \exp(E_g/2k_B T),$$

where the constant of proportionality has a weak temperature dependence (§3.5). In order to prevent the low-temperature resistances becoming too high, the semiconductor is doped with suitable impurities so that, for $kT_B \ll E_g$, the conduction is either n-type (electron carriers) or p-type (hole carriers). The details of the $\rho - T$ curve then depend on the particular semiconductor and the number and type of impurities so that an accurate representation of $R(T)$ for a thermometer is not possible. Many empirical formulae have been proposed to fit experimental values of R to a set of (calibration) temperatures T, and the normal practice is to express $\log T$ as a polynomial series in $\log R$, i.e.

$$\log R = \sum_n A_n (\log T)^n \qquad (7.10)$$

Fortunately germanium thermometers give very reproducible results, even after many cycles from 300 K to less than 1 K, and recalibration is seldom required. The resistance of a single crystal, arsenic-doped germanium thermometer (commercially available from Lake Shore Cryotronics), for example, increases from a value of 170Ω at 4 K to 840Ω at 1 K and to 21 KΩ at 0.3 K. Self-heating can sometimes be a problem and the power dissipation during measurement must therefore be kept low (for example, 10^{-9} W at 1 K, or 10^{-12} W at 100 mK). Hence a.c. techniques are frequently used.

The most commonly used (and cheapest) thermometer for work below 4 K is the standard carbon radio-resistor. As the different manufacturers employ different production processes, their characteristics vary widely and it is necessary to select a particular type and room temperature value for different temperature ranges. They are usually made up from graphite composite inside a ceramic coating and certainly do not exhibit simple semiconducting behaviour. No universal formula will describe their $R(T)$ relationship and the experimental calibration points are usually fitted to one of a number of different empirical algebraic expressions (White, 1981). A shortened version of the polynomial (7.10) is appropriate if a large enough number of calibration points is available. Carbon resistors undergo a change in calibration over a period of time but the change between successive runs gradually decreases after the resistor has been 'trained' by repeated cycling between room temperature and 4 K. An additional benefit of the carbon resistor is that it is less sensitive to magnetic fields than germanium.

A thermocouple is one of the simplest of all electrical thermometers. The junction is small in both size and heat capacity, the heat dissipation is negligible and the voltage measuring instrument provides a continuous indication of temperature. Unfortunately low-temperature thermopowers (cf. §3.3) are small—they must fall to zero according to the Third Law of Thermodynamics—and the familiar combination of copper-constantan is of little value below 20 K. Dilute gold–iron alloys, containing between 0.03 and 0.07% Fe, have the largest low-temperature thermopowers and, in combination with chromel, make thermocouples which have a sensitivity of more than $10\,\mu V/K$ at 4 K. If the reference junction is maintained at a constant temperature, AuFe thermocouples will provide a convenient and accurate measure of a temperature gradient in, for example, thermal conductivity or thermopower experiments.

Although semiconductor thermometers maintain their high sensitivity down to the lowest attainable temperatures, a means of calibrating this thermometer beyond the lowest point on the ^{3}He vapour pressure scale has to be found. Below 1 K we may look to the temperature-dependence of the magnetic properties of solids for suitable thermometric parameters. In §7.4 we noted that the magnetization of an assembly of electron or nuclear spins is a function of (B/T) and that, in the limit of a small applied magnetic field, the static susceptibility χ follows Curie's law

$$\chi = \mu_0 M/B = C/T \tag{7.11}$$

It is important to recognize that B is the *local* field B_l acting on the individual dipoles and that the difference between B_l and the applied field B_a is important in the determination of temperature. Furthermore, the simple form of Curie's law assumes that we are dealing with isolated, independent dipoles for which there is a degenerate ground state whose levels are split by an applied field. The low-field susceptibility for a paramagnetic salt with a cubic lattice can be described by the general expression

$$\chi_m = \frac{\mu_0 M}{B_a} = \chi_0 + \frac{C}{T - (\frac{1}{3} - D)C - \theta + \delta/T} \tag{7.12}$$

(Quinn and Compton, 1975) where χ_0 is included to account for any temperature-independent susceptibility.

The second term in the denominator, which is the local field correction to equation (7.11), is made up of the Lorentz dipole–dipole interaction and the shape-dependent demagnetization factor D and is zero for a spherical sample; while θ and δ/T represent exchange coupling between spins and the effect of their coupling to excited states respectively. Although in general these terms cannot be computed from first principles and can only be deduced in some special cases from spectroscopic data, prior knowledge of their values is not important since they may be found experimentally by calibration against fixed points on the thermodynamic scale. The magnetic susceptibility provides an extremely valuable secondary temperature scale because of its accuracy at temperatures well outside its calibration range. Equation 7.12 is valid down to a few times the ordering temperature of the salt, which for CMN (the most commonly used material) is about 2 mK.

The traditional mutual-inductance bridge method for measuring susceptibility has been superseded by the SQUID magnetometer, in which the changing magnetization is detected as a change in magnetic flux coupled into the primary coil of a superconducting flux transformer (§8.3). With a few mg of CMN, temperature measurements can be made down to about 5 mK in a static field of only $200\,\mu\text{T}$.

The high sensitivity of the SQUID can be used to good advantage in the measurement of the much smaller nuclear susceptibility; for a metal such as copper, it is typically 10^4 less than the electronic susceptibility of CMN under comparable conditions. Curie's law is followed with high accuracy in the limit of zero external magnetic field and the correction for finite-field values is small, so that nuclear magnetic thermometry can take over from CMN below a few mK. Again, nuclear Curie constants cannot be established by first-principle calculations and so calibration at one fixed point is necessary. A choice of methods is available for measuring a nuclear susceptibility. Static methods using SQUIDs do not involve heating of the sample, but the presence of magnetic impurities may result in an electronic contribution to the total magnetization which swamps that of the nuclear moments. Resonance, methods, using either conventional or SQUID detection, have the advantages of increased sensitivity and they can discriminate between the electron and nuclear moments, even though the former may change the resonance condition for the latter, but there may be significant eddy heating of the metallic sample. In a continuous wave system, the detector voltage is proportional to the imaginary part χ'' of the complex susceptibility. Under steady-state conditions χ'' is proportional to $\chi(B_a/\Delta B)$ where B_a is the applied field and ΔB is the width of the absorption line. Since line-widths are very narrow, that is, $< 1\,\text{mT}$, there is considerable enhancement of χ'' over the static-field value χ. In addition, heating effects can be reduced by employing a pulse technique. If the rf field is turned on for a short time, the magnetization vector, M_n, is tipped through a small angle away from its equilibrium orientation along the direction of the applied field B_a, the z-axis. The resulting transverse component of magnetization, which lies in the $x-y$ plane normal to B_a, precesses about B_a and induces an alternating voltage at the Larmor frequency in a coil whose axis lies parallel to x. This free precession signal, which decays with a characteristic time τ_2 (which is related to the line width ΔB), is proportional to the nuclear magnetization and hence to $(1/T)$. Very pure platinum is the usual material in the thermometer, and the tipping angles may be small as $4°$ in a field of less than $10\,\mathrm{mT}$. The pulsed NMR thermometer has a further advantage of being self-calibrating if the longitudinal relaxation time τ_1, which determines the recovery of the z-component of M_n after it has been perturbed from its equilibrium value, is measured, since it can be related to the absolute temperature by Korringa's relation (7.9).

We conclude this section with a brief mention of two techniques which may be used for primary thermometry at very low temperatures. Nuclear orientation thermometry depends on the anisotropic emission of γ-rays from polarized radioactive nuclei. The scale of the anisotropy in the intensity pattern is determined by the degree of polarization, which in turn depends inversely on the absolute temperature. In practice the polarization is achieved by substituting the radioactive nuclei (^{60}Co or ^{54}Mn) in a ferromagnetic host lattice so that the nuclear dipoles are aligned by the very large ($\sim 10\,\mathrm{T}$) internal field. Since the nuclear hyperfine level splittings are known with some precision, the angular distribution $W(T, \theta)$ can be computed and compared with experiment.

A rather different thermometric parameter is the Johnson or thermal noise voltage, generated by the random or Brownian motion of electrons in a resistor R. The mean square voltage measured in a detection bandwidth Δf is given by

$$\langle V_n^2 \rangle = 4 k_B T R \Delta f$$

and the main problem in adapting this equation for thermometry is in the determination of Δf. A novel way of circumventing this problem is due to Kamper et al. (1971). The resistor is incorporated in a SQUID ring containing a single Josephson junction, which is voltage-biased so as to cause the junction to oscillate at a few kHz. Noise from the resistor modulates this bias voltage and produces corresponding fluctuations in the frequency. When the spectrum is analysed, the half-power average linewidth is given by

$$\delta f = 4\pi k_B T R / \Phi_0^2 \simeq 4 \times 10^7 RT (\mathrm{Hz}).$$

Noise thermometry has the special advantage of being an absolute method and is applicable over the whole range from $20\,\mathrm{mK}$ to room temperature.

It will have become clear from the discussion in this section that the low-temperature physicist may well have to use a variety of thermometers to cover

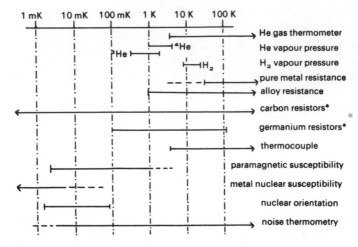

Figure 7.14 Cryogenic thermometers for use in various temperature ranges. The symbol* indicates where a single device does not cover the whole range.

a particular temperature range. Not only is this necessary for calibration purposes, but different types of measurement provide a means for checking internal consistency when there is a possibility that thermodynamic equilibrium has not been established. Cross-reference is particularly important at mK temperatures. Figure 7.14 summarizes the useful ranges of the thermometers that we have described. This review of low-temperature thermometry is by no means comprehensive and the methods described have been chosen, in part, so as to illustrate how the properties of matter and low-temperature techniques can be combined to realize accurate temperature measurement. More inclusive reviews of the topic will be found in the texts listed below for further reading.

Bibliography

References

Al'tshuler, S.A. *Zh. Eksper. Teor. Fiz. Pis'ma* **3**, 117 (1966); translated in *Sov. Phys. J.E.T.P. Lett.*, **3**, 112 (1966).

American Institute of Physics Handbook, American Institute of Physics, New York (1963).

Anderson, A.C. *Rev. Sci. Inst.* **41**, 1446 (1970).

Anufriyev, Yu D. *Zh. Eksper. Teor. Fiz. Pis'ma* **1**, 1 (1965); translated in *Sov. Phys. J.E.T.P. Lett.* **1**, 155 (1965).

Bradley, D.I., Bradshaw, T.W., Guénault, A.M., Keith, V., Locke-Scobie, B.G., Miller, I.E., Pickett, G.R. and Pratt, W.P. Jr. *Cryogenics* **22**, 296 (1982).

Campbell, S.J., Herbert, I.R., Warwick, C.B. and Woodgate, J.M. *J. Phys. E. (Sci. Instruments)* **9**, 443 (1976).

Hedgecock, F.T. and Muir, W.B. *Rev Sci. Inst.* **31**, 390 (1960).

Kamper, R.A., Siegworth, J.D., Radebaugh, R., Zimmerman, J.E. *IEEE Proc.* **59**, 1368 (1971).

Keesom, W.H. *Helium*. Elsevier, Amsterdam (1942).

Korringa, J. *Physica* **16**, 601 (1950).
Little, W.A. *Physica* **109 B & C**, 2001 (1982).
Mate, C.F., Harris-Lowe, R., Davis, W.L. and Daunt, J.G. *Rev. Sci. Instr.* **36**, 369 (1965).
Mueller, R.M., Buchal, C., Folle, H.R., Kubota, M. and Pobell, F. *Cryogenics* **20**, 395 (1980).
Pomeranchuk, I. *Zh. Eksp. Terr. Fiz.* **20**, 919 (1950).
Quinn, T.J. and Compton, J.P. *Rep. Prog. Phys.* **38**, 151 (1975).
Rubin, L.G., Brandt, B.L. and Sample, H.H. *Cryogenics* **22**, 491 (1982).
Sydoriak, S.G., Roberts, T.R. and Sherman, R.H. *J. Res. NBS* **68A**, 559 (1964).
Van Dijk, H. *Prog. Cryogenics* **2**, 123 (1960).
Walton, D. *Rev. Sci. Inst.* **37**, 734 (1966).
Zemanski, M.W. and Dittmann, R.H. *Heat and Thermodynamics*. McGraw-Hill, New York (1981).
Zijlstra, H. *Experimental Methods in Magnetism*. North Holland, Amsterdam (1967).

Further reading

Betts, D.S. *Refrigeration and Thermometry below One Kelvin*. Sussex University Press, London (1976).
Hoare, F.E., Jackson, L.C. and Kurti, N. *Experimental Cryophysics*. Butterworths, London (1961).
Lounasmaa, O. *Experimental Principles and Methods below 1 K*. Academic Press, London (1974).
Rose-Innes, A.C. *Low Temperature Laboratory Techniques*. 2nd edn., English Universities Press, London (1973).
Rosenberg, H.M. *Low Temperature Solid State Physics*. Clarendon Press, Oxford (1963).
Swenson, C.A. 'Physics at high pressure', in *Solid State Physics* **11**, 41 (1960).
Webb, A.W., Gubser, D.U. and Towle, L.C. 'Cryostat for generating pressures to 100 kilobar and temperatures to 0.03 K', in *Rev. Sci. Inst.* **47**, 59 (1976).
White, G.K. *Experimental Techniques in Low Temperature Physics*. 3rd edn., Clarendon Press, Oxford (1979).

8 Applications

8.1 Uses of low temperatures

As we pointed out in chapter 1, there is no clear-cut upper limit to what might be termed a low temperature. Low-temperature applications might start with a domestic refrigerator working at a few degrees below the ambient room temperature. To comply with the general spirit of this book, the effects to be described in this chapter will be restricted by an upper limit of the normal boiling point of liquid air, and the major part will deal with phenomena associated with much lower temperatures. We are here concerned with how the particular physical properties of matter at low temperatures may be used in or applied to other branches of science, including engineering and medicine. For this reason the applications of different aspects of superconductivity will occupy much of the subsequent discussion. These are described in three sections: high-current and high-magnetic-field technology; application of the Josephson effect; and a brief review of other small-scale applications. In the final section of this chapter we will review how the fundamental properties of the liquid gases can be utilized in a variety of areas outside the research laboratory.

There are, of course, many processes, experiments and devices which can only be operated at temperatures below 100 K. In other situations performance and efficiency are improved by cooling. Since biological and chemical activity virtually ceases at low temperatures, the preservation of, for example, food, human tissues, blood and bone marrow becomes possible by careful cooling to temperatures near 77 K. Cell tissue can also be permanently destroyed by sudden exposure to very cold liquids, and the removal of defective tissue, both internal and external, has been achieved by the localized application of liquid nitrogen. Cryo-biology, -medicine and -surgery are 'growth areas', and the miniature liquid-nitrogen refrigerator could become a standard piece of hospital equipment, although the hospital physicist is more likely to be concerned with the application of the low-temperature properties of inanimate objects to medical diagnosis, e.g. superconducting magnets for NMR scanners, SQUID magnetocardiography, etc.

It is in the field of electronics, however, particularly at high frequencies, where the benefits of cooling are likely to be most extensive. These benefits are not confined to superconducting devices and circuits. The limiting factor of any receiver system used in the detection of very weak electromagnetic signals is the noise generated by the primary detecting or mixing element and the following amplifier. Since the noise in most devices is determined by the random, thermal motion of current carriers in resistive elements, known as Johnson noise, improved performance may be expected if the operating temperature is reduced. However, not all semiconductor devices will function in the cryogenic environment. In both n-type silicon and germanium, $k_B T$ falls below the ionization energy as the temperature is reduced and, if the operation of a device depends on the number of thermally excited carriers, carrier freeze-out will degrade performance. In n-type III-V compounds, such as GaAs, the impurity band overlaps the valence band even for low levels of impurity concentration and freeze-out does not occur. Both JFETs and MOSFETs have been used at 4 K. In GaAs JFETs increased carrier mobility leads to greater transconductance (g_m). As the dominant thermal noise in the channel varies as T/g_m, cooling an amplifier gives a higher gain and a lower effective noise temperature. Varactors, tunnel diodes and Zener diodes will all work at low temperatures as long as there is no carrier freeze-out. A noise figure T_N of less than 10 K has been reported for a single-stage, helium-cooled, 430 MHz varactor-tuned amplifier (Prance et al., 1982). GaAs FET amplifiers working up to frequencies of 10 GHz, a tunnel-diode oscillator tunable over the range 200–300 MHz for NMR and CMOS integrated circuitry have all been successfully operated at cryogenic temperatures.

Improved detection of radiation from the microwave region up to the far infrared may be achieved by cooling semiconductor devices such as broadband detectors (including bolometers) and heterodyne mixers. Comparisons with the equivalent Josephson device will be made in a later section, but some details of a particular device are worth noting here. Cooled infra-red detectors are regularly employed in space and upper-atmosphere experiments. The presence of the background cosmic radiation, with a black-body distribution centred on 3 K (peak wavelength 1.7 mm), has been detected by measuring the emission spectra of the night sky using a cooled spectrometer. A two-stage bolometer, which consists of a thin film absorber coupled to an InSb temperature sensor, operating at $T = 0.35$ K in a charcoal-pumped ^{3}He cryostat, was the detector. The telescope was flown on a balloon platform at an altitude of 43 km. Expressed as a noise equivalent power, the minimum detectable signal for a signal-to-noise-ratio of unity was 6×10^{-16} W in a recording system with 1 Hz bandwidth (Richards, 1981).

8.2 High-current and magnetic-field applications of superconductivity

The fundamental property of a superconductor, that a steady current will flow in a wire without dissipation, offers the possibility of increased efficiency in situations where large currents are involved. Nowhere is the potential benefit more obvious than in electric power technology where the system power exceeds 10 MVA, for example, in the alternator of a power station or a power transmission line, or when the stored energy is greater than 10^7 J, for example, in the magnets for a large bubble chamber or a magneto-hydrodynamic generator. Most of these systems involve high current densities and high magnetic fields, both of which will eventually restore a superconductor to the normal or resistive state. The search for materials which retain their superconducting properties under these extreme conditions has centred on the type II alloys and compounds of the transition metals niobium and vanadium, since these have amongst the highest values of the upper critical fields H_{c2} and also the highest transition temperatures (see §4.3). The refrigeration problems that are (and will be) encountered are very different from those of the laboratory, and one of the major problems is the maintenance of a continuous supply of refrigerants in systems which may extend over several miles. In this section we will concentrate on the development of the materials that are now incorporated in magnets, motors, alternators and transmission lines.

In §4.3, we saw that the penetration of flux into a type II superconductor begins at a relatively low field intensity $\mu_0 H_{c1}$ but that the last trace of superconductivity is not eliminated until the much higher field $\mu_0 H_{c2}$ is applied. In the intervening field range, the mixed-state, flux quanta thread this material, forming a regular 2-D lattice structure which becomes gradually more tightly packed as the field is increased. The reversible magnetization curve of Fig. 4.5 can be explained by the work of Abrikosov, but a type II material which followed this curve would be characterized by a critical current density which falls to zero close to H_{c1} and would therefore be of little use in high-field applications. When a transport current J_T flows in a superconductor which is in the mixed state, a Lorentz force acts on the flux lines so as to move them at right angles to their axes and to the direction of current flow. Movement of magnetic flux causes a voltage to be induced across the sample in the same direction as the current flow. Thus the material becomes resistive and heat is dissipated. The motion can be prevented, or at least suppressed, however, if the flux lines are pinned by defects in the crystal structure. The critical current density of a type II superconductor, which may be defined as that current above which there is continuous movement of flux lines or at which a measurable voltage can be detected, is therefore a function of the metallurgical state of the material. Resistanceless operation above H_{c1} is achieved only in *hard* superconductors where the pinning forces are large and which, consequently, are characterized by highly irreversible magnetization curves. Many kinds of defects will pin the flux lines, including dislocations, grain boundaries, vacancies and clusters of impurities. Magnetization and current-density versus applied magnetic field curves are shown in Fig. 8.1 for a material with an increasing level of work hardening.

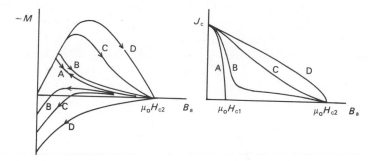

Figure 8.1 Magnetization and critical current versus applied magnetic field for samples of the same material with increasing amounts of flux pinning (A → D).

Two materials in particular have been predominantly involved with high-field and high-current superconducting technology. These are the alloy of niobium and titanium ($Nb_{40}Ti_{60}$) and the compound Nb_3Sn. Other compounds with the same crystalline structure (A15) have been investigated but the relatively small differences between their properties and the metallurgical problems associated with producing a commercial conductor are such that the present discussion can be limited to Nb_3Sn. In the alloy, flux lines are pinned by tangles of dislocations which form during the drawing of the wire and, in Nb_3Sn, the highest critical current is obtained in material with the smallest grain size, indicating that grain boundaries are the cause of flux pinning. The phase diagrams for the two materials are shown in Fig. 8.2. Although their T_c values are greater than 10 K (the minimum working temperature for H_2), the reduction of critical current with increasing temperature means that most applications require the use of liquid helium. Their metallurgical properties are actually quite different: the alloy is strong and ductile and can be readily drawn down in large quantities to a diameter of about 0.2 mm; while by contrast, Nb_3Sn is brittle and weak, and the fabrication of a conductor in a form suitable for winding into magnet coils presents a number of major technical problems. However, the design of cables is determined primarily by factors other than metallurgical ones.

The first superconducting magnets in the 1960s were found to have performances which were not in accordance with measurements made on short samples of the same wire. A coil would revert to the normal, resistive state, at a quarter of its expected current value in a process which is known as quenching. Repeated excitation or 'training' of the magnet would often lead to an improvement. It was recognized that some precautions were necessary to prevent a local quench (caused by a small region of the winding exceeding the critical current density) from propagating and causing a full-scale quench, with the possible destruction of the magnet insulation through very large induced voltages. A simple solution to this problem is to clad the super-

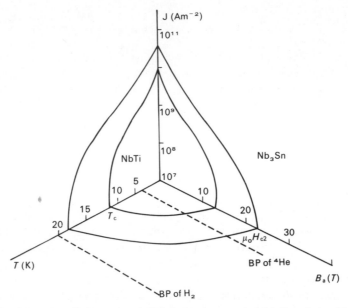

Figure 8.2 Phase diagrams for two type II superconductors, showing the relationship between critical current, magnetic field and temperature.

conductor with a low-resistance normal conductor, such as copper. This forms a temporary alternative path for the current in the event of a local resistive region being formed in the superconductor. Composite conductors consisting of strands of superconductor embedded in a matrix of copper have been successfully used for the partial stabilization of small magnets carrying up to 100 A but, for larger magnets, additional cryostatic stabilization is required. The ratio of copper to superconductor in the windings is greatly increased (typically 20:1 or more) and helium is channelled throughout the winding, so as to be able to remove the heat generated by a local quench without a catastrophic rise in temperature. These fully-stabilized conductors have the disadvantage that the current density is reduced to about $10^7 \, Am^{-2}$.

An alternative solution to this problem was found when one of the sources of the instabilities that cause premature quenching was understood. As the current in a coil is increased, flux lines will penetrate the superconducting regions and although pinning provides resistance to motion, penetration into the bulk must occur for increasing magnetic field. However, under certain conditions the closed current loops around the flux lines become unstable and collapse spontaneously. The subsequent motion of the flux, called a flux jump, causes local heating and with it a significant a rise of temperature, since the heat capacity and the thermal conductivity are both low. Increasing temperature reduces the critical current and the pinning force: further flux jumps are then produced and a flux avalanche may develop.* However redistribution of flux causes a reduction in the flux gradient and in the Lorentz force driving the flux motion.

Since the temperature rise is related to the dimensions of the superconducting element, the flux avalanche may be avoided if the reduction in this Lorentz force is greater than the reduction in the pinning force. The condition for stability is that the conductor diameter d should be smaller than a value given by

$$d^2 < 8C/(\mu_0 J_c |dJ_c/dT|)$$

where C is the conductor heat capacity and J_c is the required critical current density.

Thus it is possible to reduce the effect of flux jump instabilities by replacing a single conductor with a very large number (up to 10^5) of superconducting filaments with diameters of about $10\,\mu m$ embedded in a matrix of copper or aluminium. These superconductors clearly combine a number of stabilization effects: cryogenic stabilization, filamentary stabilization and also dynamic stabilization, since the presence of a pure metal with low electrical resistance will slow down the propagation of magnetic disturbances. For the filamentary stabilization to be effective, however, the filaments must be independent of one another, insofar as there is no current flow between them through the normal matrix. This condition is satisfied in steady-state operation but when the conductor is subject to a changing field, either in a.c. operation or when turning the device on or off, an induced e.m.f. causes current flow in the matrix. Twisting the filaments about the axis of the wire helps to reduce the magnetic coupling: if the necessary degree of twist cannot be achieved, however, a barrier of high resistance cupro-nickel is incorporated in the copper matrix. Finally hysteresis losses in the individual superconducting filaments are minimized by further reduction in diameter to about $5\,\mu m$.

Thus the final composition and structure of a high-current superconducting cable is somewhat complicated. Fabrication of these composites presents additional problems, particularly in the case of Nb_3Sn which, because of the brittleness already noted, cannot be drawn down from a billet to the mm dimensions of the final wire. An ingenious solution is to fabricate a wire starting from rods of pure Nb in a matrix of bronze (an alloy of copper and tin) which are ductile, and then subsequently to use a heat treatment which causes Sn to diffuse out of the bronze and into the Nb, where it reacts to form Nb_3Sn. It is also possible to wind a magnet coil with the untreated wire if the epoxy

---

* In the critical-state model of an irreversible type II superconductor (Bean, 1962), it is assumed that the current at any point is either equal to the critical current value or zero. Flux lines are everywhere in a state of dynamic equilibrium, with the Lorentz force tending to produce flux motion being balanced by a pinning force. Flux penetration proceeds through a sequence of critical states.

resin which is used for potting can withstand the subsequent heat treatment at about 700°C.

The simplest application of this superconducting technology is in the construction of coils for generating constant or slowly-varying magnetic fields. Their use was restricted initially to the research laboratory but an increasing number are now replacing conventional electromagnets in standard laboratory equipment, e.g. in magnetic spectrometers, and in industrial applications, e.g. magnetic filtration and purification. In the future the efficient operation of storage systems, magnetohydrodynamic generators and thermonuclear fusion reactors, will rely on superconductors. Indeed, fusion reactors which use magnetic confinement, e.g. tokamaks, will require superconducting magnets if the power consumption is not to exceed the reactor's output.

Magnets can be divided into two broad categories: small or laboratory magnets are those which use intrinsically stable (multifilamentary) cable, while large magnets are those in which full cryostatic stabilization is required, to provide adequate protection for the windings in situations where the stored energy can be greater than 10^8 J. In the former the aim is to achieve high fields (up to 17T) over relatively small volumes (typically 10^{-4} m^3), with high current densities (10^8–10^9 Am^{-2}) but employing modest power supplies. Economic factors are not usually a primary consideration. Various configurations, such as a single coil, Helmholtz pairs, or nested coils are used to provide the desired field strength and homogeneity. Epoxy impregnation (potting) helps to reduce conductor movement and hence frictional heating as well as providing a means for containing the magnetic forces. If a superconducting switch is mounted across the coil's terminals, the magnet may be run in a persistent mode by shorting out the current leads from the power supply once the required current has been established. Such a switch consists, essentially, of a length of superconductor which can be driven into the normal state, thus 'opening' it, either by means of a small heater or by the fringing field of the magnet itself. Cooling by immersion in liquid ^{4}He at 4 K is standard practice but higher critical current densities in NbTi coils may be achieved by cooling to 1.2 K.

In large magnets, overall volume is not a primary limitation and conductors with high copper to superconductor ratios are operated at lower current densities (10^7 Am^{-2} in the case of NbTi composites) to give maximum fields of between 5 and 10 T. Most of these large magnets are to be found in nuclear-physics laboratories around the world. The bubble-chamber magnet at CERN has a working volume of 80 m^3 and a current of 6000 A can create a field of 3.5 T with a stored energy of 800 MJ, while the 1000 GeV proton accelerator at the Fermilab (Chicago) uses nearly 1000 superconducting magnets to steer and focus the beam along its 6.4 km path. Problems which face designers of these massive projects include: (1) containment of stresses, both magnetic and thermal, in complex winding configurations; (2) the maintenance of the correct position and alignment of magnets over great distances in accelerators; (3) maintaining an adequate supply of refrigerant, usually in the form of supercritical ^{4}He at c. 4.5 K, which can be forced to flow through the windings or through the conductors themselves if these are hollow; and (4) protection of magnets, leads and power supplies as well as the safe dissipation of stored energy in the event of a quench.

The benefits of using superconductors in electrical machinery have yet to be fully realized. A number of motors and generators have been constructed and these have demonstrated the feasibility of incorporating superconducting components. Future development will depend more on economic and reliability factors rather than on the improvement of present technology. Applications where size, weight and efficiency are of prime importance would appear to be most promising. There is one obvious complication with all electrical machinery; some of its parts must rotate at very high speeds. In the conventional turbine-driven alternator, a direct-current electromagnet is rotated within the static (armature) coils from which the a.c. power is drawn. Reversing the position of field windings and the armature would be advantageous for cryogenic and engineering reasons but would create severe electrical problems in removing power from the rotor. High a.c. losses at power frequencies (even in filamentary conductors) have restricted the use of superconductors to the field windings and even then they must be protected from the alternating fields created by the current in the normal metal static windings. The shield, which forms part of the rotor, reduces the armature-field coupling and degrades performance. In spite of the electrical, mechanical and cryogenic problems, the energy losses of a conventional alternator could be reduced by up to 60% with the use of superconductors, and the reduced size and weight of such a superconducting system would clearly be extremely attractive for airborne operations.

In a homopolar machine, the stationary field coils are not subject to a torque reaction nor to time-varying fields produced by the armature current. Thus d.c. motors and generators using superconducting windings are of the homopolar type, rather than commutated heteropolar machines. The rotating disc or drum is made from a normal conductor and the main problems have been associated with current transfer to and from moving contacts. Interest has centred particularly on the development of motors for ship propulsion: prototype systems with a mechanical output of greater than 1000 h.p. have been constructed, and motors with power an order of magnitude greater are under development.

Finally, we should note that under certain conditions the efficiency of power transmission from generator to consumer could be improved by the use of superconducting cables. Material problems are not serious since the magnetic fields are much smaller, typically < 0.1 T, and suitable conductors have been fabricated and tested successfully for transmission of 10^9 VA. However, the anticipated gain in efficiency was only achieved for power levels greater than present demand and meanwhile there are comparable advantages from using other, less extreme, forms of cooled conductors.

We have been able to give only a superficial account of type II materials and the devices that are fabricated from them in this important area of superconducting technology. The reader wanting to explore the subject in greater depth is directed towards the books listed at the end of this chapter.

8.3 Applications of the Josephson effects

Of all the properties of superconductors, those associated with the Josephson effect provide the most diverse range of technical applications across the whole spectrum of science, engineering and medicine. In this section we can provide only an overview of the subject under the three broad headings; SQUIDs; high-frequency applications; and computer elements. Many excellent texts are available to the reader who wants further information, and these are listed at the end of this chapter.

SQUIDs

In §4.6 it was shown that a parallel configuration of two Josephson junctions in a superconducting loop behaved like an interferometer, insofar as the critical current for the device (that is, the maximum current that flows without a voltage appearing across it) oscillates with a period of Φ_0 when subjected to an increasing applied magnetic flux. Since loop areas can be made quite large ($\simeq 1 \, \text{cm}^2$), the magnetic field periodicity may be as small as $10^{-11} \, \text{T}$, and thus the interferometer can form the basis of extraordinarily sensitive measuring instruments for magnetic and electrical effects. The acronym SQUID (Superconducting quantum interference device) is given to a variety of devices which have the common property that their response to applied flux is oscillatory with period Φ_0. The simplest SQUID consists of a single Josephson junction in a superconducting ring which is coupled inductively to an external circuit. No d.c. current is fed to the ring but in this case the circulating current is an oscillatory function of the applied flux. Because the SQUID ring response is detected through the behaviour of the coupled r.f. tuned circuit, the single-junction device is referred to as an r.f. SQUID. Both the r.f. SQUID and the double-junction d.c. SQUID can be made from the three types of weak link described earlier (§4.6), but the most successful devices have used niobium point contacts in the first type and thin-film tunnel junctions in the second. Before describing the uses of SQUIDs we will give a brief outline of their different modes of operation. As was noted before, the depth of the critical current modulation ΔI in the d.c. SQUID depends on the amount of screening produced by the circulating current, which in turn depends on the product LI_0, where I_0 is the critical current for each junction and L is the ring inductance. The choice of both parameters is limited. L must be kept small to reduce the effects of thermal noise, but not so small as to limit flux-input coupling. To avoid hysteresis in the weak links, I_0 must also be kept small, but again there is a lower limit determined by the size of ΔI and the dynamic

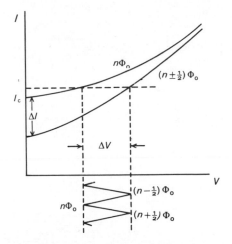

Figure 8.3 Voltage variation across a resistively shunted d.c. SQUID as the flux through the loop is changed.

resistance of the SQUID, which falls with decreasing I_0. For these reasons, a value of $LI_0/\Phi_0 \simeq 0.5$ is chosen and the modulation depth is then about 50%, or $\Phi_0/2L$. The device is usually operated in the finite voltage regime with $I > I_c$. Modulation of I_c leads to a change in the shape of the $I-V$ characteristic for the SQUID as shown in Fig. 8.3. For a fixed bias current fed into the ring, the voltage across it is modulated as the external magnetic flux is changed, by an amount approximately equal to the dynamic resistance R_D times the critical current modulation, i.e.

$$\Delta V \simeq R_D \Delta I_0 = R_D \frac{\Phi_0}{2L}. \tag{8.1}$$

The variation of ΔV with applied flux is approximately triangular and thus the flux sensitivity, $\Delta V/\Delta \Phi \simeq R_D/L$.

The basic circuit for an r.f. SQUID is shown in Fig. 8.4. The tuned circuit (sometimes called the tank circuit) is fed from a current source at an r.f. frequency of about 20 MHz and the effect of the inductively coupled, superconducting circuit is to produce a load which varies periodically with the applied (d.c.) flux. To understand how the device works, we need to find a

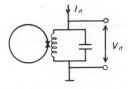

Figure 8.4 Basic circuit for an r.f. SQUID.

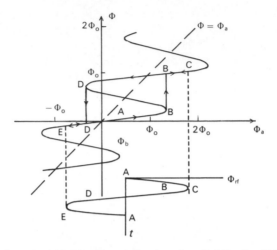

Figure 8.5   Flux through an r.f. SQUID as a function of applied flux for $LI/\Phi_0 \geqslant \tfrac{1}{2}\pi$.

relationship between the flux linking the superconducting ring and the applied flux Φ_a. This can be achieved using the London equation (4.7) and the Josephson equation (4.24). The flux through the ring can be expressed as

$$\Phi = \Phi_a - LI \tag{8.2}$$

where I is the circulating screening current and L is the SQUID ring inductance, and the phase difference across the junction is related to the flux Φ by

$$I = I_0 \sin\left(2\pi\Phi/\Phi_0\right). \tag{8.3}$$

The coupled equations may be solved to find Φ in terms of Φ_a but the behaviour depends again on the parameter $LI_0/\Phi_0 = \beta$. For $\beta \ll 1$, there is no screening and $\Phi = \Phi_a$, while for $\beta \gg 1$ the screening is complete and $\Phi = 0$ for all values of Φ_a. For values of $\beta \gtrsim \tfrac{1}{2}\pi$, the $\Phi v. \Phi_a$ curve has regions with negative slope as shown in Fig. 8.5. These regions are unstable and for increasing (or decreasing) applied flux, the coupled flux Φ will only follow those parts of the curve with a positive slope.

Let us consider how the flux Φ changes during one cycle of the r.f. current which produces the sinusoidal variation Φ_{rf} as shown in Fig. 8.5. We will also assume that there is a small background d.c. flux Φ_b, which is not entirely screened by the circulating current I. This corresponds to the point A in the figure. As the r.f. flux increases from zero, the flux through the loop and the circulating current will increase. At the point B, the slope $(d\Phi/d\Phi_a)$ is infinite and the system becomes unstable. The weak link behaves like a gate which opens to allow a single flux quantum Φ_0 to enter the loop during the upward transition from B to B'. The circulating current also changes discontinuously and reverses its direction. Between B and C, the flux Φ and the current I once more increase. As Φ_{rf} is reduced, the path is not reversed and the flux quantum is not expelled from the circuit until the point D is reached, when there is a downward transition in $\Phi(D \rightarrow D')$ and another reversal of I. For the remainder of the cycle, the flux follows the reversible path along the lower branch.

The net effect of taking the SQUID around the hysteresis loop is to cause energy dissipation and a reduction in the effective Q-factor of the tank circuit. This is a threshold effect. As the r.f. current bias increases, the amplitude of the oscillating voltage across the tuned circuit is linear in I_{rf} while the amplitude of Φ_{rf} is less than a value corresponding to the point B in the cycle of Fig. 8.5. The onset of hysteresis reduces the level of oscillation in the tuned circuit and the time average of the voltage $\langle V \rangle$ remains constant, until the r.f. current has been increased sufficiently to overcome the additional loss in the secondary superconducting circuit. The resulting $\langle V \rangle - I_{rf}$ characteristic is a staircase pattern, with a series of plateau regions which occur every time Φ_{rf} is increased sufficiently to cause the SQUID to go round an additional hysteresis loop. The value of the threshold voltage and current for each step will clearly depend on the magnitude of the static flux Φ_b coupled into the ring. If the SQUID is biased at the middle of one of the steps and the value of the applied d.c. flux is varied, the value of $\langle V \rangle$ will oscillate with an approximately triangular waveform of period Φ_0, as shown in Fig. 8.6.

Thus both the d.c. and r.f. SQUIDs give voltage outputs which are approximately triangular in the applied flux. In the simplest mode of operation for a magnetometer, magnetic flux, and hence magnetic field, can be measured by counting the number of periods of V that result from a change in the applied field. Greater sensitivity is achieved using a feedback method. A low frequency flux signal Φ_m is fed into the SQUID ring., either by a separate coil in the case of the d.c. SQUID or by providing amplitude modulation to I_{rf} in the r.f. SQUID. If the SQUID is set at a turning point, such as P in Fig. 8.6, the flux modulation produces no voltage output at the modulation frequency. If the bias is then changed to P' as the result of an external flux, there will be a component of voltage V_m at the modulation frequency. This signal is amplified and phase-sensitively detected to give a d.c. output which is a direct measure of

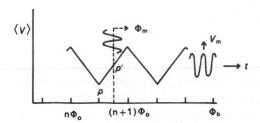

Figure 8.6 Amplitude of the voltage across the tank circuit of an r.f. SQUID as a function of the steady flux Φ_b. A similar flux variation is detected as the voltage across a d.c. SQUID.

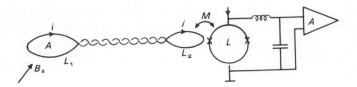

Figure 8.7 Superconducting flux transformer for coupling flux into a SQUID.

the flux difference $(\Phi_{P'} - \Phi_P)$ and one which may be used as a feedback signal to bring the bias point back to P.

Flux from an external source can be coupled directly into a SQUID but, because of its high sensitivity, it is frequently used to measure small changes in magnetic field (e.g. the the signal from a human heartbeat). The SQUID is then screened from other external sources (using a superconducting enclosure) and the flux to be measured is coupled into the ring via a superconducting flux transformer. In its simplest form, this consists of two coils which are connected by a pair of closely twisted leads as in Fig. 8.7. One coil (L_1) samples the local magnetic field while the second (L_2) couples flux into the SQUID. For the total flux through the transformer to remain constant, the applied flux in L_1 must be compensated by an induced current (i) flowing through both coils, such that

$$\Phi_a = (L_1 + L_2)i. \tag{8.4}$$

The flux coupled into the SQUID will depend on the mutual inductance M between L_1 and the SQUID ring. Thus

$$\Phi_{\text{SQUID}} = \Phi_a M/(L_1 + L_2). \tag{8.5}$$

By careful choice of coil parameters, flux gain can be achieved in certain circumstances.

A variety of configurations using different Josephson weak links have been successfully employed in the design of SQUIDs, and a limited number are available commercially. Of these the two-hole r.f. SQUID has been most widely used. This is an adjustable point contact device, machined from a cylinder of solid niobium, which has the advantage of self-shielding, since the response is determined by the difference in the flux coupled through the two holes. The coil of the resonant tank circuit is inserted in one hole and the sensing coil (e.g. L_2 of the flux transformer) is inserted in the other. A toroidal version of this SQUID provides even better shielding. Because of the problems associated with the adjustment of two matched junctions, the most successful d.c. SQUIDs have been fabricated with oxide barrier tunnel junctions, made from niobium, lead or a lead-niobium combination. Early versions used a three-dimensional structure, e.g. the SQUID was deposited on the outer surface of a cylindrical quartz tube (see Fig. 8.8) to ensure tight coupling between the SQUID and the input coil, but more recently planar structures have been successfully developed.

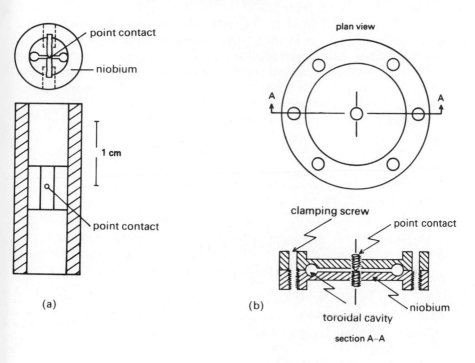

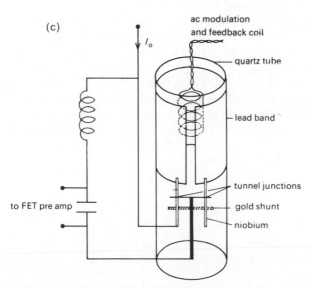

Figure 8.8 Practical r.f. and d.c. SQUIDS. (*a*) Two-hole point contact and (*b*) toroidal structured point contact r.f. SQUIDS (Zimmerman *et al.*, 1970; Rifkin *et al.*, 1976). (*c*) Cylindrical d.c. SQUID showing the output circuit and modulation coil (Clarke *et al.*, 1975).

The ultimate sensitivity of any SQUID system is limited by random noise voltages, generated either by the SQUID ring or by the associated electronic amplifiers. It is usually expressed as an equivalent flux noise in a bandwidth of 1 Hz at the input to the detection system. For a typical r.f. SQUID operating at 4 K and 20 MHz, the flux resolution is about $10^{-4}\,\Phi_0\,\mathrm{Hz}^{-\frac{1}{2}}$. Rather better figures have been achieved with d.c. SQUIDs.

An alternative figure of merit is the energy in an input coil carrying a current δi which can just be resolved by the SQUID, i.e.

$$\varepsilon = \tfrac{1}{2} L_2 (\delta i)^2 \tag{8.6}$$

This minimum current is limited by the flux noise, which, when expressed as a spectral density S_Φ, is related to δi by

$$(S_\Phi)^{\frac{1}{2}} = M\,\delta I.$$

Thus the energy resolution of any SQUID becomes

$$\varepsilon = S_\Phi / 2\alpha^2 L \qquad \mathrm{JHz}^{-1} \tag{8.7}$$

where α is the coupling constant defined by $M = \alpha \sqrt{L_2 L}$. For d.c. SQUIDs, a figure of merit of about $1 \times 10^{-33}\,\mathrm{JHz}^{-1}$ (or $\sim 2h$) has been reported, but typical commercial r.f. and practical d.c. SQUIDs have values of between 10^{-28} and $10^{-29}\,\mathrm{JHz}^{-1}$. A comprehensive survey of various types of SQUIDs is given in a review article by Clarke (1977), and more recent advances in d.c. SQUIDs have been described by Ketchen (1980).

SQUIDs are employed in many areas of science and medicine. Although the SQUID is essentially a flux meter, it may be converted into a meter for measuring current, voltage or resistance, or into a null detector, by a suitable choice of circuitry connected to the sensing coil L_2. The d.c. current sensitivity is typically better than $10^{-9}\,\mathrm{A}$, with an extremely low input impedance, and this number can be improved by increasing the number of turns on the input coil. For potentiometric applications, the SQUID is operated in a flux-locked loop with the output being fed back to the input circuit as shown in Fig. 8.9. A resolution of better than $10^{-12}\,\mathrm{V}$ can be achieved with a dynamic range of about $10^{-6}\,\mathrm{V}$ and an input impedance of $10\,\Omega$: although this last figure is small

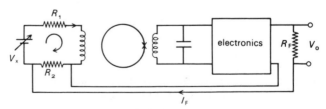

Figure 8.9 Circuit for measuring an unknown voltage V_x. The current I_1 generates flux in the SQUID and the electronics feeds back a current V_0/R_F such that $I_F R_2 = V_x$. At balance $I_1 = 0$ and $V_x = V_0 (R_F/R_2)$.

by room-temperature standards, the impedances in the low-temperature circuit are usually even lower.

In their role as magnetometers SQUIDs are used to measure susceptibilities of very weakly magnetic materials, and in low magnetic fields, their sensitivity is at least one order of magnitude better than that of conventional methods, such as NMR. In geophysics, the SQUID magnetometer or, more usually, gradiometer has been used to detect fluctuations in the earth's magnetic field. Perhaps the most spectacular application of SQUIDs is in the detection of the magnetic fields produced by the human body which are of the order of 10^{-11} T or smaller. Heart activity (magnetocardiograms) and brain activity (encephalograms) are both recorded by SQUID systems, often in an environment which does not involve elaborate shielding against external sources of magnetic field. This remarkable achievement is made possible by the use of flux transformer coils which are only sensitive to second or higher derivatives of magnetic fields. In the area of fundamental research, SQUID sensors are being used in the search for gravitational waves and magnetic monopoles.

High-frequency applications

Because the d.c. characteristics of Josephson junctions are modified by temperature and electromagnetic radiation, their value as low temperature sensors has been fully realized in the last few years. There are a variety of ways in which they may be used but we will limit the discussion to broad-band detectors, including bolometers and mixers in the microwave and the near-infrared frequency bands. A bolometer is a thermal detector which yields an output in direct proportion to the rise in temperature caused by the absorption of radiation. It is a two-component device with a radiation-absorbing element (usually a thin bismuth film deposited on a sapphire substrate) and a resistance thermometer, and its use is limited to the infra-red, since single photon detectors in the visible region and heterodyne detectors in the microwave bands have greater sensitivities and faster response times. Three temperature sensors offer comparable cryogenic performance: an SNS Josephson junction (e.g. a normal metal sandwich Pb/Cu–Al/Pb) which employs the strong temperature-dependence of the critical current $I_0(T)$.

$$I_0(T) = I_0(0) \exp(- T/T_0)^{\frac{1}{2}}; \tag{8.8}$$

a transition edge bolometer using the temperature-induced resistance change in a superconducting film biased at T_c: and a helium-cooled semiconductor of the sort described in §8.1. All three types have a sensitivity (expressed as a noise equivalent power NEP) of about 10^{-15} W in a 1 Hz detection bandwidth, but the semiconductor has the shortest time constant of 10^{-2} s.

Like a conventional diode, the Josephson detector follows a square-law response to small a.c. signals, i.e. the output voltage is proportional to the

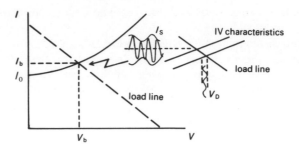

Figure 8.10   Use of a Josephson junction as a broad-band detector. The inset shows the modulation of the $I-V$ characteristic by an applied a.c. signal. (After Richards *et al.*, 1973.)

input power. Its operation depends on the deformation of the $I-V$ character-istic to an amplitude modulated signal, as shown in Fig. 8.10. If the junction is biased at a constant current slightly larger than I_0, the output voltage follows the input signal modulation and its magnitude at the modulation frequency is given by

$$V_D = \gamma I_s^2 R_D I_0 \tag{8.9}$$

where γ is a constant which determines the change in the maximum (zero voltage) current I_0 to the a.c. current I_s, and R_D is the dynamic resistance of the detector at the bias point. The detector's performance may be compared with that of a derivative of the Schottky barrier diode, which has a superconductor-semiconductor junction. Enhanced low-temperature properties of the super-Schottky diode result, in part, from the disparity between the superconducting and semiconducting energy gaps which differ by a factor of about 1000. A super-Schottky diode, cooled to 1 K, detects 10 GHz radiation with an NEP of 2×10^{-15} W/Hz, while a Josephson junction at 4 K, detecting 90 GHz radiation, has an NEP of 5×10^{-15} W/Hz.

In heterodyne detection, the incoming small signal is mixed in a non-linear device with the output from a local oscillator (LO) at a frequency close to that of the signal. The output at the difference or intermediate frequency (IF), is amplified in a narrow-band amplifier prior to the video detection which recovers the amplitude modulation of the original signal. Since the Josephson junction is itself a source of radiation when it is voltage biased, it may be used as a self-oscillating mixer, in which the external signal is mixed with the fundamental Josephson frequency v_J or with one of the harmonics. Although this arrangement has the attraction of simplicity, there are problems asso-ciated with stabilizing v_J against internal fluctuations and variations in the bias voltage. Thus the mixer is usually operated with an external LO so that the IF signal does not depend explicity on v_J. For a signal I_s that is small compared to the LO current, and small IF, the current in the junction at the

LO frequency is amplitude-modulated at the IF by an amount determined by I_s. As the amplitude of the LO-induced current step is a function of the LO current amplitude, the $I-V$ characteristic is modulated at the IF. For a current-biased junction,

$$V_{IF} = \frac{R_D I_s}{2} \frac{\partial I_0}{\partial I_{LO}} \tag{8.10}$$

Two figures of merit characterize the performance of any mixer. These are the conversion effeciency η (usually a loss), which is the ratio of the power coupled into the IF amplifier to the available signal power, and the mixer noise temperature, so that the overall receiver noise temperature, T_R, is given by

$$T_R = T_m + T_{IF}/\eta \tag{8.11}$$

where T_m and T_{IF} are the noise temperatures of the mixer and IF amplifier respectively. A value of η greater than unity has been obtained in a Josephson mixer detecting a 36 GHz signal with a measured T_m of 54 K (Claasen et al., 1974). Super Schottky diodes are found to have lower T_m values, but with values of $\eta < 1$.

Detection and fundamental or harmonic mixing by Josephson devices can be extended up to frequencies in excess of 1000 GHz ($\lambda < 0.3$ mm), while high-order (> 100) harmonic mixing of microwave oscillators with infrared laser signals provides an accurate measure of laser frequency. Finally it is worth noting that a Josephson junction can form the basis for parametric amplifiers and tunable radiation sources, although in the case of the latter, the highest measured power at 10 GHz is about 10^{-9} W.

This section concludes with a brief discussion of how a Josephson junction is used for the determination of the fundamental constant e/h and as a laboratory voltage standard. When a junction is irradiated with microwave radiation, a sequence of constant voltage steps is induced in its $I-V$ characteristic (Fig. 4.19). Under the influence of an applied voltage, the current through the junction has an oscillatory component with frequency

$$v_J = 2eV/h. \tag{8.12}$$

This oscillating current will phase-lock to the applied radiation or to the harmonics generated by the non-linear response of the junction and the voltage is held constant for a range of current which may be as wide as several mA. The position of the nth step is given by

$$V_n = n \cdot \frac{h}{2e} v \tag{8.13}$$

where v is now the frequency of radiation. For $v = 10$ GHz, $V_1 \simeq 20\,\mu$V. Since the frequency of a source can be determined with a high degree of accuracy (1 part in 10^8 or better), a measurement of the step separation (by comparison with the voltages from standard cells) yields a direct measure of the fundamental constant e/h and in turn a value of the fine-structure constant $\alpha(= e^2/4\pi\varepsilon_0 \hbar c)$ which is independent of any assumptions of quantum electrodynamics. By adopting a new value for $2e/h$ (equal to $4.835\,944 \times 10^{14}\,\text{Hz}\,\text{V}^{-1}$), the international volt is now defined in terms of the Josephson frequency. A variety of experiments on different materials, types of junction, temperature, frequency and step number indicate that the fundamental relationship for v_J is exact to at least 1 part in 10^8.

Superconducting computer elements

The search for even faster computers has led scientists to look at new technologies for computer design. Cryogenic computers, which make use of the unique properties of superconductors, have been investigated for a number of years and it is now possible to construct an instrument in which all of the basic elements, including the logic circuits to process information, the memory stores and the interconnecting lines are superconducting, and which runs about 50 times faster than the conventional alternatives. The fundamental advantage of going to low temperatures is that high-speed switching can be achieved in a Josephson tunnel junction with very low power dissipation. A simple gate and its I–V characteristic are shown in Fig. 8.11. A single junction is driven into the resistive state ($V \neq 0$) by raising the gate current I_g above the critical current value I_0 for the weak link or by reducing I_0 below I_g by the application of a magnetic field. For switching by magnetic induction, the junction is overlaid by an insulated thin-film control line so that the field produced by I_{CL} penetrates the barrier region. In the $V \neq 0$ state the junction resistance is much higher than the load resistor R_L and the device switches along the load line so that I_g is almost totally steered to the output line. To achieve high sensitivity to the control current, the area of the junction is increased but the benefit is offset by an increase in switching time, which is partly determined by the capacitance of the barrier region. This problem is overcome by using interferometers containing two or more junctions as the switching element. There is then inductive coupling between the control line and the superconducting strips which make up the arms of the interferometer. The gate circuit in Fig. 8.12 contains three junctions with critical currents in the ratio 1:2:1 and the threshold curve for $I_g - I_{CL}$ is equivalent to a three-slit

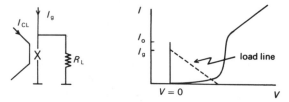

Figure 8.11   A simple Josephson gate with its switching characteristic.

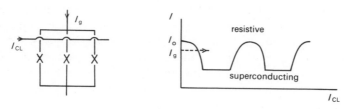

Figure 8.12 Schematic of a 3-junction interferometer and the threshold curve for switching to the resistive state.

interference pattern. Such a device can have a switching time as short as 20 ps.

One problem associated with the switch is its latching property. Once the device is in the $V \neq 0$ state it remains there after the control current is reduced to zero. A return to the $V = 0$ state is only obtained by removing the current I_g: this is readily achieved, however, if the computer's bipolar clock supplies current to the gate. Unlike a semiconductor device, the Josephson gate is not sensitive to the direction of current flow and it will function with either polarity of current, resetting itself as I_g passes through zero.

Logic elements such as OR and AND gates can be constructed from combinations of gates and control lines. Figure 8.13 gives a schematic representation of a logic OR gate. The left-hand interferometer has two control lines so that switching occurs when either is activated. The gate current, which is switched into the output line, is used to control one or more interferometers in a fan-out process. If the interconnecting transmission lines have the correct impedance matching, logic delays of less than 50 ps are possible. In the AND gate there are two interferometers in parallel, each with its own control line, and an output voltage occurs only if both control lines are activated together.

Superconducting memory makes use of Josephson switches but also of flux quantization in a superconducting loop. Information is stored as trapped flux by causing a persistent circulating current to flow in the loop. The binary states 1 and 0 may be differentiated by the sense of the current or by using zero current for one of them. A basic version of a memory cell (Fig. 8.14) consists of a superconducting loop with a Josephson gate in one of its arms. The state of

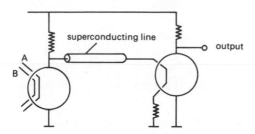

Figure 8.13 Schematic representation of an OR logic circuit.

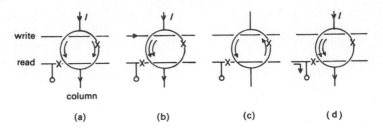

Figure 8.14 A simple superconducting memory element, showing how a bit of information is 'written' and 'read'.

the gate is controlled by a 'write' line, while the state of a second gate, which lies under one arm of the loop, in the 'read' line, is controlled by the loop current. To store a bit of information, binary 1, currents are supplied simultaneously to the loop (acting as the column address in an array) and to the 'write' line. The loop current, which in the absence of the write signal would divide equally between the two arms, is switched into the left branch when the Josephson junction in the right branch is driven into the resistive state by the write current. Removal of first the write current and then the column line current (by driving the same current through the loop in the opposite direction), causes a circulating current to be trapped in the loop, which has returned to the superconducting state. The read operation requires current to be fed to both loop and read lines. If the persistent current is present, the additional current in the left branch is sufficient to switch the Josephson junction in the read line and the read current is directed into the output line.

These memories have a number of attractive properties: zero power is dissipated during storage, the switching time is about 0.1 ns, very little energy ($\sim 10^{-18}$ J) is required in switching from one state to another if a small number of flux quanta are trapped, and the memory is non-volatile while the circuit remains superconducting. In the device described above, the read-out is also non-destructive.

Cryogenic computers make use of the fabrication technology (such as photolithography) which was developed for semiconductor microcircuits. Typically devices are fabricated from conductors with dimensions of 2.5 μm and 0.2 μm, and high-density packing allows up to 1000 gates to be sited on a 6 mm $\times$ 6 mm chip. As the dissipation is less than 1 μW/gate (the voltage and currents involved are $\simeq$ 1 mV and $\simeq$ 1 mA respectively), the dissipation at the limiting packing density for conventional circuitry is easily handled by superfluid helium. Thus the necessary cooling powers of the helium liquefier for a M-byte instrument is typically a modest 10 W.

8.4 Other small-scale applications of superconductivity

Much of the early technology involving superconductors has since been superseded by the Josephson technology described in the previous section.

The first thin-film superconducting switch, the cryotron, used the current in the control line to provide a field greater than the H_c value of a continuous gate line. Properties of type I superconductors, other than those associated with Josephson effects, are however a source of useful devices in applications which do not involve large currents.

First we will look at some simple applications based on fundamental properties. The very rapid fall in resistance at the transition to the super-conducting state is used in the edge bolometer mentioned in the last section and the well-defined T_c values in pure materials provide convenient fixed points on a secondary thermometry scale. The properties of flux transfor-mers, which are widely used in the input circuits to SQUID and flux-gate magnetometers, are based on the Meissner effect, as is the magnetic screening which is afforded by a superconducting enclosure. A constant magnetic field can be trapped in a superconducting cylinder if the source of the field is removed after the metal has been cooled through T_c.

One of the experimental techniques outlined in §2.5 uses a tunnel junction as a phonon generator. When quasiparticles are generated as a result of pair breaking (§4.5), they are injected into states above the energy gap. This is a non-equilibrium situation and the quasiparticles relax first to the gap edge, where the density of states $D(\epsilon)$ is high, and they then recombine as pairs. Phonons are emitted in both steps and their spectrum is consequently made up of two contributions: there is a peak at a frequency $v = 2\Delta/h$ corresponding to the recombination of two quasiparticles, while the continuous background (shown in Fig. 2.12) results from the relaxation of quasiparticles to the gap edge. By modulating the junction bias voltage and using phase-sensitive detection of the emitted phonons, the junction becomes a tunable and almost monochromatic source. A simple spectrometer for the frequency band 10^{10}–10^{12} Hz can be fabricated by depositing tunnel junctions on opposite faces of a test sample. When either recombination or relaxation phonons with energy $hv > 2\Delta_D$, where Δ_D is the energy gap for the second (detector) junction, are absorbed, pair breaking increases the number of quasiparticles. Thus the tunnelling current at a detector bias voltage $V < 2\Delta_D/e$ will be enhanced over its thermal value by phonon absorption. A comprehensive review, which includes details of fabrication and deployment of the rather specialized tunnel junctions required for phonon spectroscopy, is given by Eisenmenger (1976).

A very recent development (Faris, 1983) exploits the non-equilibrium properties of superconductors (§4.7) in conjunction with quasiparticle tunnel-ling in a two-junction sandwich (called a QUITERON) which has a transistor-like switching characteristic and which therefore provides current gain.

Losses in high-frequency transmission lines made from normal metal conductors are usually acceptable except, however, in the input lines to low-noise amplifiers and in cases where very narrow pulses are transmitted and long distances are involved. A superconducting parallel plate strip line, fabricated from Nb and PbIn conductors and separated by a 0.1 μm thick dielectric layer, has been designed for the transmission of ns pulses with an attenuation at $10\,\mathrm{GHz}$ of $10^{-4}\,\mathrm{dBm}^{-1}$ at $T = 1\,\mathrm{K}$, and would, for example, form the connecting lines in superconducting computers.

The quality (Q) factor of a microwave cavity is limited by the surface resistance of its walls. The improvement that might in principle be expected from using superconductors is not easily realized. For a pure conductor, the resistance can be expressed by

$$R_s = \frac{C\omega^{3/2}}{T}\exp\left(\frac{-\Delta(T)}{k_B T}\right) + R_0(\omega) \qquad (8.14)$$

and it is the second term, the residual resistance, that restricts the Q-factor. Trapped flux, impurities and surface finish contribute to R_0 while small signal Q values exceed those obtained under the high power conditions that are found (say) in the reasonators of linear accelerators. A typical small-scale application is in oscillator stabilization, where Q values of 10^{11} at $10\,\mathrm{GHz}$ can be achieved. This is a factor of 10^6 greater than the Q values of a copper resonator operating at room temperature.

8.5 Uses of liquefied gases

In this section, we will consider very briefly a few of the practical applications that depend on the properties of cryogenic refrigerants. An obvious but none-the-less valuable property of all liquid is that for a given mass, the volume in the liquid phase is considerably less (a factor of 10^{-3} in the case of $^4\mathrm{He}$) than that in the gaseous phase, and the liquid state is often therefore preferred for the mass transport and storage of materials which would be gaseous at room temperature. For example in the steel industry, where very large quantities of oxygen are used to improve furnace efficiency and to purify the molten metal, many mills have their own air-liquefaction and oxygen-separation plants, which produce several hundreds of tons of liquid oxygen per day. Rockets carry their fuel (hydrogen), and the oxygen to support combustion, in liquid form; and the large quantities of helium that are required by the oil industry as an additive to the oxygen supplied to deep-sea divers (a typical mixture for deep diving being $2\%\ O_2$, 98% He) are transported by air and road in liquid form.

The quality of vacua in laboratory and industrial systems is improved by using liquid N_2 cold traps and by cryopumping with liquid $^4\mathrm{He}$. The principle of both operations is the same. When the walls of the vacuum chamber are cooled to the temperature of a surrounding cryogenic fluid, molecules of residual gases in the chamber will condense onto the cold surfaces if their boiling points are higher than that of the refrigerant. As the vapour pressure of all substances (apart from the two isotopes of helium) is negligibly small at $4\,\mathrm{K}$, the vacuum space between the walls of a helium dewar will be maintained at an extremely low pressure by cryopumping—as long as helium itself is excluded.

The superfluid properties of $^4\mathrm{He}$ have some rather special uses. In an acoustic microscope, the coupling liquid between the sapphire lens and the

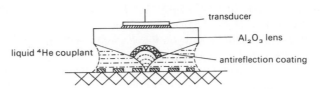

Figure 8.15 Acoustic microscope lens system which uses liquid ⁴He as the couplant (after Heiserman *et al.*, 1980).

object must have very low loss at high frequency if the maximum benefit of the diffraction-limited resolution is to be realized. Most liquids have an acoustic attenuation which increases as the square of the frequency but, at temperatures below the λ-point, the attenuation in ⁴He falls rapidly and at 4 GHz and 0.1 K, its value is a factor of 10^3 less than that of water at the same frequency and at room temperature. A scanning microscope using superfluid ⁴He coupling at 100mK has been used by Quate and co-workers (Heiserman *et al.*, 1980) to produce micrographs of the electrode in a transistor which have a resolution comparable to that of an SEM but which also show additional details of the strain in the surface resulting from ion implantation: details of the lens are shown in Fig. 8.15.

Naturally-occurring helium gas trapped above oil deposits is not isotopically pure ⁴He but it includes about 2 parts in 10^7 of ³He. For some experiments and applications this level of impurity is unacceptable, but ⁴He purification can readily be effected using its superfluid properties below T_λ. One method is to use a heat flush (see §6.5). The ³He impurity forms part of the normal fluid component at $T < T_\lambda$ and, if the liquid is heated, the normal and superfluid components move in opposite directions in the temperature gradient. The process is shown schematically in Fig. 8.16. Using an apparatus based on this principle, McClintock (1978) achieved an impurity level of better than 1 part in 10^{15}.

Ultrapure superfluid ⁴He may have an important function in experiments designed to determine whether the neutron has an electric dipole moment. Large numbers of ultra-cold neutrons (with energies of *c*. 10^{-7} eV) are required. Cold neutrons from a reactor can be down scattered to the required extremely low energy by the creation of phonons in HeII. Because the resultant ultra-cold neutrons are reflected specularly by the walls of the helium container, they become trapped in what is called a neutron bottle (Golub,

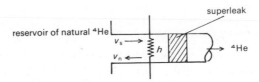

Figure 8.16 Isotopic purification of ⁴He by a 'heat flush'. Normal fluid flows away from the heater (*h*) with velocity v_n and the superfluid flows towards it with velocity v_s, before passing through a superleak to the collection vessel.

1983). In order that the neutrons should not be lost through absorption by ^{3}He, the ratio of ^{3}He atoms to ^{4}He atoms must be less than 1 part of ^{3}He per 10^{11}, a purity level which is, of course, readily achievable using the heat flush technique described above.

Cryophysics is a relatively new science and the technology that it is now developing from it is still in its infancy. It is inevitable that there will usually be a lag between the discovery of new phenomena in the research laboratory and their wider application to commercial and industrial systems. There can be no doubt, however, that significant technological advances in many different areas will follow in due course from the further exploitation of the properties of matter at low temperatures.

Bibliography

References

Bean, C.P. *Phys. Rev. Lett.* **8**, 250 (1962)
Claasen, J.H., Taur, Y. and Richards, P.L. *Appl. Phys. Lett.* **25**, 759 (1974).
Clarke, J., Goubau, W.M. and Ketchen, M.B. *Appl. Phys. Lett.* **27**, 155 (1975).
Faris, S.M. *Phys. Bulletin* **34**, 420 (1983).
Golub, R., Jewell, C., Ageron, P., Mampe, W., Heckel, B. and Kilvington, L. *Z. für Phys. B (Condensed Matter)* **51**, 187 (1983).
Heiserman, J.E., Rugar, D. and Quate, C.F. *J. Acoust. Soc. Am.* **67**, 1629 (1980).
Ketchen, M.B. *IEEE Trans. on Magnetics* **MAG17**, 387 (1981).
McClintock, P.V.E. *Cryogenics* **18**, 201 (1978)
Prance, R.J., Long, A.P., Clark, T.D. and Goodall, F. *J. Phys. E* **15**, 101 (1982)
Richards, P.L. Auracher, F. and Van Duzer, T. *Proc. IEEE* **61**, 35 (1973).
Richards, P.L. *Phil. Trans. Roy. Soc.* **307**, 77 (1982). See also: Woody. D.P., Nishioka, N.S. and Richards, P.L. *J. de Physique* **C6**, 1629 (1978).
Rifkin, R., Vincent, D.A., Deaver, B.S. and Hansma, P.K. *J. Appl. Phys.* **47**, 2645 (1976)
Zimmerman, J.E., Theine, P. and Hardening, J.T. *J. Appl. Phys.* **41**, 1572 (1970).

Further reading

Adde, R. and Vernet, G. 'High frequency properties and applications of Josephson junctions from microwaves to far-infrared' in *Superconductor Applications: SQUIDs and Machines*, Ed. Schwartz, B.B. and Foner, S., Plenum, New York (1976).
Clarke, J. 'Superconducting Quantum Interference Devices for Low Frequency Measurements', ibid.
Barone, A. and Paterno, G. *Physics and Applications of the Josephson Effect*. Wiley, New York (1982).
Deaver, B.S., Falco, C.M., Harris, J.H. and Wolf, S.A. *Future trends in Superconductive Electronics*. Am. Inst. of Phys., New York (1978).
Eisenmenger, W. 'Superconducting tunnelling junctions as phonon generators and detectors', in *Physical Acoustics*, Ed. Mason, W.P. and Thurston, R.N., vol. 14 (1976), Academic Press, New York.
Lengeler, B. 'Semiconductor Devices Suitable for Use in a Cryogenic Environment'. *Cryogenics* **14**, 439 (1974).
Luhman, T. and Dew-Hughes, D. 'Metallurgy of superconducting materials' in *Treatise on Material Science and Technology*, vol. 14, Academic Press, New York (1979).
Van Duzer, T. and Turner, C.W. *Principles of Superconductive Devices and Circuits*. Arnold, London (1981).

Index

Authors whose works are cited are referred to in the Bibliographies at the end of each chapter.

'ff' denotes that the discussion of a topic continues through succeeding pages; **boldface** entries represent the location of the central discussion of a topic.